Wired Wilderness

ANIMALS, HISTORY, CULTURE

Harriet Ritvo, *Series Editor*

Wired Wilderness

*Technologies of Tracking and
the Making of Modern Wildlife*

Etienne Benson

The Johns Hopkins University Press
Baltimore

© 2010 The Johns Hopkins University Press
All rights reserved. Published 2010
Printed in the United States of America on acid-free paper
9 8 7 6 5 4 3 2 1

The Johns Hopkins University Press
2715 North Charles Street
Baltimore, Maryland 21218-4363
www.press.jhu.edu

Library of Congress Cataloging-in-Publication Data

Benson, Etienne, 1976–
   Wired wilderness : technologies of tracking and the making of
modern wildlife / Etienne Benson.
      p. cm.
   Includes bibliographical references and index.
   ISBN-13: 978-0-8018-9710-8 (hardcover : alk. paper)
   ISBN-10: 0-8018-9710-6 (hardcover : alk. paper)
   1. Animal radio tracking—Moral and ethical aspects. 2. Animal
radio tracking—Technological innovations. 3. Conservation biology—
Environmental aspects. I. Title.
   QL60.4.B46 2010
   639.9—dc22   2010001246

A catalog record for this book is available from the British Library.

*Special discounts are available for bulk purchases of this book. For more
information, please contact Special Sales at 410-516-6936 or
specialsales@press.jhu.edu.*

The Johns Hopkins University Press uses environmentally friendly
book materials, including recycled text paper that is composed of
at least 30 percent post-consumer waste, whenever possible. All of our
book papers are acid-free, and our jackets and covers are printed on
paper with recycled content.

Contents

Acknowledgments

Most of the research for this book was completed while I was a student in MIT's Doctoral Program in History, Anthropology, and Science, Technology, and Society. During that period and since, Harriet Ritvo has been a thoughtful, rigorous, and generous advisor and friend. Her scholarship on human-animal relationships profoundly shaped this book, as it has many others. I am also indebted to Meg Jacobs and Stefan Helmreich, the other members of my dissertation committee, for their support and constructive criticism over the course of this project.

Financial support for this project was generously provided by a variety of organizations. In addition to stipendiary support from the HASTS Program, MIT's Kenan Sahin Presidential Fellowship, the Martin Family Foundation, and Resources for the Future, I received research and travel funding from the HASTS Program, the MIT Center for International Studies, the MIT Kelly-Douglas Fund, the Society for the History of Technology, and the National Science Foundation. The revision of the book manuscript was completed during a two-year postdoctoral fellowship from the Harvard University Center for the Environment. The Harvard Department of the History of Science and my faculty host Janet Browne provided a welcoming institutional home during the fellowship period.

In the course of this project many scientists, engineers, and conservationists answered my phone calls and e-mails, gave me tours of their field sites and laboratories, provided photographs and other documentation of their work, and submitted to more or less formal interviews. Even those whose names do not appear in the final version of the book provided insights that helped shape the work as a whole. I thank Robin Baird, Timothy Beaulieu, Jean Bourassa, Harry Brower Jr., Marianna Childress and the staff of Collecte Localisation Satellites, William W. Cochran, Lance Craighead, Frederick C. Dean, Jeff Foster, Philippe

Gaspar, Charlotte Girard, M. Bradley Hanson, Bridget Kenward and the staff of Biotrack, Robert E. Kenward, Valerian B. "Larry" Kuechle and the staff of Advanced Telemetry Systems, Clarence Lehman and the staff of the Cedar Creek Natural History Area, Rexford D. Lord Jr., Greg Marshall, L. David Mech, Hemanta R. Mishra, Christian Ortega, Arlo Raim, Philippe Roques, Naomi Rose, Glen and Beverly Sanderson and their daughter Laurie Sanderson, John C. Seidensticker, Donald B. Siniff, George Swenson, Michel Taillade, John R. Tester, Fernando Ugarte, Dwain W. Warner, and William Woodward.

The assistance of numerous archivists and librarians was vital to this project, particularly those of the American Museum of Natural History in New York; the Bibliothèque Centrale of the Muséum National d'Histoire Naturelle in Paris; the Denver Public Library; the U.S. Department of Agriculture's National Wildlife Research Center in Fort Collins, Colorado; the U.S. Fish and Wildlife Service's National Conservation Training Center in Shepherdstown, West Virginia; the Ernst Mayr Library at Harvard University's Museum of Comparative Zoology; the Illinois Natural History Society in Champaign, Illinois; the Minnesota Historical Society in St. Paul; Montana State University in Bozeman; the Massachusetts Institute of Technology; the National Archives and Records Administration in College Park, Maryland, and Anchorage, Alaska; the National Marine Fisheries Service's Southwest Fisheries Science Center in La Jolla, California; NMFS's National Marine Mammal Laboratory in Seattle, Washington; the Patuxent Wildlife Research Center in Patuxent, Maryland; the Scripps Institution of Oceanography in La Jolla, California; the Smithsonian Institution in Washington, DC; the Tuzzy Consortium Library in Barrow, Alaska; the University of Alaska, Anchorage; the University of Alaska, Fairbanks; the University of California, Berkeley; the University of Montana, Missoula; the University of Minnesota; the University of Washington; the University of Wisconsin; the University of Wyoming; the Woods Hole Oceanographic Institution in Woods Hole, Massachusetts; and the Yellowstone National Park Heritage Research Center in Gardiner, Montana.

Over the course of this book's long gestation, many colleagues and friends have provided helpful comments, critiques, and conversations. I am especially grateful to Peter Alagona, Mark Barrow, Alexandre Benson, Laurel Braitman, Nicholas Buchanan, Candis Callison, Kieran Downes, Deborah Fitzgerald, Xaq Frohlich, Bernard Geoghegan, Shane Hamilton, Sabine Höhler, David Kaiser, Eben Kirksey, Shekhar Krishnan, Vincent Lépinay, Thomas Levenson, Leo Marx, Lisa Messeri, Natasha Myers, Sophia Roosth, Ryan Shapiro, Hanna

Rose Shell, Jenny Leigh Smith, Michaela Jane Thompson, William Turkel, Nick Wilding, Michael Wise, Rebecca Jane Woods, Sara Ann Wylie, and Anya Zilberstein.

Finally, I am grateful to family and friends in Minneapolis, San Diego, San Francisco, Seattle, Washington, DC, and elsewhere who offered me their spare beds and couches, lent me their cars, cooked me delicious meals, accompanied me on hikes in grizzly country, and otherwise put me deeply in their debt in the course of this project. Most of all, I would like to thank my parents, Evelyne and William Perry Benson.

Wired Wilderness

Knowing the Wild

Many Americans in the second half of the twentieth century were fascinated with wild animals. They watched wildlife films and television shows, visited zoos, aquariums, and amusement parks with performing wild animals, donated money to organizations working to "save" baby seals, whales, pandas, tigers, and other charismatic creatures, and gave their support to politicians who promised to protect wild animals and their habitats, sometimes even at the cost of economic growth. They valued national parks and wilderness areas as much for the bears, wolves, elk, and other animals inhabiting them as for their scenic vistas or dramatic geological formations, and they fell in love with the raptors and other once-threatened species that began recolonizing urban areas once legal protections were in place. Conservationists and scientists learned to frame their concerns about habitat loss, pollution, and climate change in terms of the threats they posed to wild animals, recognizing that reports of the possible sighting of an ivory-billed woodpecker or the image of a polar bear on the edge of a melting ice floe were often more effective ways of stimulating action than statistics about annual rates of deforestation or rising atmospheric carbon dioxide levels.

This fascination with and concern for wild animals supported a boom in wildlife research. Even as the proportion of Americans who hunted wild animals for pleasure or profit shrank, undermining the constituency that had largely supported wildlife research and conservation from the late nineteenth century to the mid-twentieth century, new sources of support grew. The federal environmental legislation passed in the years around the first Earth Day in 1970—especially the National Environmental Policy Act of 1969, the Marine Mammal Protection Act of 1972, and the Endangered Species Act of 1973—evinced a widespread suspicion toward narratives of modernity and progress, but it also enthroned science and technology as the most promising means of mitigating the effect on wild animals of growing human populations and levels of consumption. Scientists, after all, had often been the first to sound the alarm about vanishing wildlife, and their knowledge and expertise seemed indispensible to the project of allowing a diversity of living things and habitats to coexist with humanity.[1]

This faith in and support for science stimulated a search for more effective ways of studying often-elusive wild animals in their natural habitats. Often this search was framed in terms of what the environmental historian Gregg Mitman has called a "transcendent vision" of nature, which would make it possible to restore a lost, Edenic nature. Of these techniques, none had such a dramatic impact on the everyday practice of wildlife biologists or inspired so many encomiums to the potential for technology to "save nature" as wildlife radio tracking or radiotelemetry. Originating around 1960 at the unlikely intersection of wildlife management and military surveillance technologies, the use of miniaturized radio tags and collars to keep track of individual animals became virtually a sine qua non of wildlife research by the 1980s, dominating the pages of professional publications such as the *Journal of Wildlife Management* and serving as a symbol of modern wildlife conservation for observers of the field. One historian writing in the late 1980s described "the wolf with the radio collar, providing data for scientists to use in reestablishing the primitive ecosystems of North America," as "the perfect symbol of our efforts to come to terms with our knowledge of nature's order, our power over it, and our need to preserve our mythic past." Another, a historian of big game hunting in the British Empire, described the radio tagging of a rhinoceros in Nepal as "the perfect symbol for the replacement of the hunting by the conservation ethos, imperial power by post-colonial environmental concerns." Wedding Americans' fascination with the wild to their equally fervent enthusiasm for technology, the rise of radio tracking as the privileged mode of knowing wild animals seems both ironic and inevitable.[2]

Such is the story that can be read in the existing histories of modern wildlife

conservation and in the accounts of leading conservationists and wildlife biologists. There is another, less well-known story, however, that can only be pieced
together from archival sources, oral histories, and scattered news reports. This
alternative story reveals fractures within the seemingly perfect, if ironic, mar-
riage of Americans' interest in wildlife and in science and technology. Through
these fractures a very different, much messier, and far more conflict-ridden history of the role of science in modern wildlife conservation becomes visible. As
this book shows, technologies of wildlife research were the focus of a long-
running, pervasive debate within the community of those interested in wildlife conservation, if "community" can be used to describe such a varied and
sometimes tenuously connected network. Bound together by a shared interest
in conserving wild animals, this community was internally fractured by deep
differences over the very meaning and value of "wildlife"—differences that were
reflected in their opinions about wildlife radiotelemetry.

Why, after all, did so many Americans care about wild animals? Was it because they hoped to preserve a vanishing frontier experience that they believed
was essential to the American national character, as Teddy Roosevelt and other
sportsman-conservationists of the late nineteenth and early twentieth century
had? Because they saw wild animals as windows onto evolutionary or ecological processes, whose loss would forever compromise our ability to understand
the natural world, as many twentieth-century scientists did? Because they saw
wildlife as essential elements of complex ecosystems upon which the health and
survival of all living creatures depended, as many late-twentieth-century conservationists did? Or because they shared the humane concerns of nineteenth-
century advocates of animal welfare or the more radical animal rights philoso-
phies of the late twentieth century, which attributed inherent, inalienable value
to each individual animal life, whether or not it was a member of the human
species? Privileging one or the other of these reasons for valuing wildlife could
lead to very different conclusions about the proper means for "saving" it, and
apparently superficial debates over means forced supposed allies to confront
profound differences over ends.

In telling this alternative story of conflict and contestation over the practices
of wildlife biology, this book builds on recent developments in several subfields
of historical scholarship, particularly environmental history and the history of
science and technology. Since the early 1990s, environmental historians have
been grappling with challenges to received ideas of wilderness, most notably
expressed in William Cronon's much-debated essay "The Trouble with Wil-

derness," which argued that wilderness is a problematic human construct rather than simply a natural object. The stories told in this book reveal a richer and more nuanced discourse about the meaning of wilderness and wildness in the twentieth century than either the supporters of this argument or their critics have tended to recognize. The wilderness absolutism they critique or defend was only one thread within a broader tapestry, some of whose most vivid and illuminating scenes depict disagreements over the proper means of studying and managing "wilderness wildlife." This book also builds on recent scholarship concerning the political, cultural, and social values inherent in the practices and material culture of scientists. By focusing on an applied field science, wildlife biology, that attracted the interest and concern of nonscientists of various kinds, it shows that late-twentieth-century science was less closed to "public engagement"—a misleading euphemism for what were often adversarial contests driven both by differences in fundamental values and by mutual incomprehension—than is often assumed.[3]

The argument that wildlife conservation and the science that supports it are contentious and politicized is, of course, not new. American wildlife managers and biologists have been complaining about "biopolitics"—understood as political interference into decisions properly left to experts—since at least as far back as the 1930s, when they first established the journals, conferences, professional associations, degree programs, and financial supporters that allowed them to lay claim to the status of an autonomous, self-accrediting profession. Conservation activists have regularly protested the manipulation of policy by (other) special interests. New administrations in Washington have brought sudden reversals in supposedly science-based government policies; populations designated as "threatened" or "endangered" under the Endangered Species Act have been delisted under one administration only to be relisted under the next, with little if any change in the scientific evidence. This sort of political conflict is well worth attending to, but as this book argues, disputes over the interpretation and application of scientific findings are not the only or, in many cases, the most important way in which wildlife biology becomes imbued with social values. As the history of wildlife radiotelemetry over the past half century shows, an engaged public, consisting often of small but highly vocal activists, some of them also scientists, has shaped the techniques that scientists can use and thus the kinds of findings that may be politicized in the first place.

CHAPTER ONE

Cold War Game

In search of new methods of locating, tracking, and identifying individual wild animals in their natural habitats, wildlife biologists in the 1950s looked to the electronic technologies of the Cold War and the space age. The recent invention of the transistor had made it possible to build, for the first time, radio transmitters small enough to be attached to living creatures without causing significant changes in behavior. One of the first biologists to seize upon the opportunity presented by this technological development was Dwain W. Warner, an ornithologist at the University of Minnesota. Born in 1917, Warner had grown up on a dairy farm near Northfield, Minnesota, the home of Carleton College, where he completed an undergraduate degree in botany and zoology in 1939. He began his graduate work in ornithology in 1939 at Cornell University under Arthur Allen, a pioneer in field recording of birdsongs. Warner was initially interested in studying Mexican birds, but when he was posted to the South Pacific as part of a biological team during the Second World War, he shifted his focus to that region's birds for his dissertation. After finishing his degree in 1947, Warner was hired as an assistant professor and curator of birds at the University of Minnesota's Museum of Natural History. There his research interests returned to Mexico, many of whose migratory birds passed through Minnesota on their way to and from northern breeding sites.[1]

At Cornell Warner had learned techniques for collecting, banding, and observing birds that served as the foundation of his first decade of research in Minnesota. By the mid-1950s, he had also begun exploring the possibility of applying recent advances in instrumentation in the physical sciences to ornithology. This relatively vague set of interests was transformed into a concrete research program by the Sputnik launches of the fall of 1957, the second of which carried a live dog. Warner recalled sharing his frustration with Athelstan Spilhaus, dean of the Institute of Technology at the University of Minnesota and inventor of the bathythermograph, an important tool for oceanography and submarine warfare: "The Russians put the dog Sputniks up there, and they're having telemetered back to the Earth both physiological and ecological data on the capsule, and physiological . . . data on the dogs. . . . We haven't even done that out the window here."[2]

With Spilhaus's support, Warner began seeking funding for a bioelectronics laboratory at the Museum of Natural History. At the same time, he sought a research site where he could deploy animal tracking devices as well as a variety of environmental sensors. In March 1958, Warner described how as-yet untapped resources in the "technological world" would make possible "concise, accurate, and continuous measurements" of ecological systems to Arthur N. Wilcox, director of the University of Minnesota's Cedar Creek Natural History Area. Newly developed telemetric technology would soon make it "possible to have radioed into the laboratory on the campus and recorded on tape and film through every minute of the 365 days of the year numerous kinds of data on micro- and macroclimate, phenology, movements of animals both large and small, even predation and other mortality factors." "It seems logical," Warner added, "that, if man is eventually (even soon!) to obtain direct data on environments on moon and planets, the equipment should first be tested on earth. And, if man, other animals and plants are to be transported to other planets, a more thorough understanding of the inter-relationships is imperative." Warner pointed out that the funds necessary to adapt the instrumentation of the physical sciences to field biology were "staggering to the imagination of the biologist, but, fortunately, are not so to the technologists."[3]

Unlike a park or a wildlife refuge, Cedar Creek was devoted solely to natural history and ecological research. In the 1930s the Minnesota Academy of Science had acquired the initial parcels of land and convinced the University of

Minnesota to take responsibility for managing the site as a research reserve. In 1940, Wilcox, then chairman of the Academy's Committee on the Preservation of Natural Conditions, noted that the Academy was "primarily interested in seeing this forest preserved as nearly as possible in the undisturbed condition so that its original natural history values will be maintained." That meant that the use of the site for research or education should "be in the nature of observation rather than experimentation." Still, Wilcox suggested, there was some former crop and pasture land within the reserve where "careful restoration" to original ecological conditions would be appropriate, and there were agricultural fields that had only recently been abandoned or were still being cultivated that Wilcox suggested could serve as sites for limited "experimental work." In 1958, at the dedication of Cedar Creek's new laboratory building, Theodore Blegen, dean of the University of Minnesota's graduate school, took pains to emphasize the site's strictly scientific mission: "Cedar Creek is not a park. It is not a picnic grounds. It is not a tourist center. It is not a recreation spot for those of us who have tired nerves. It is an outdoor laboratory, a place for study, a natural terrain for biological scholars."[4]

Warner took advantage of the building of new infrastructure at Cedar Creek in 1958 to advance his vision of comprehensive environmental surveillance. As part of a collaborative application to the National Science Foundation, Warner proposed connecting environmental sensors in Cedar Creek's woods and fields to the new laboratory building by a network of buried electrical cables. John R. Tester, who had recently earned his Ph.D. at the University of Minnesota under wildlife biologist William H. Marshall with a dissertation on wildlife management in northwestern Minnesota's prairies, helped Warner seek out the proper instrumentation. In October 1958, Wilcox informed Warner that NSF had come through with a grant of $55,800, which would cover, among other things, Warner's electrical cables and environmental sensors.[5]

Meanwhile, Warner's enthusiasm for what he called "biotechnology" or "biophysics"—the application of the physical sciences and engineering to field biology—helped launch another project under the leadership of Marshall. In December 1957, the regional office of the U.S. Fish and Wildlife Service in Minneapolis had begun discussions with the Minneapolis-Honeywell Regulator Company, which was involved in defense and space-related projects, about applying the company's "electronics know-how" to wildlife management. The Ordnance Division of Honeywell's Seattle Development Laboratory had recently built a sonic tag for tracking salmon on the Columbia River. Although

Warner was not directly involved in these discussions, his interest in using miniature radio transmitters to track wildlife inspired the Fish and Wildlife Service
staff to ask Honeywell if such a device was possible.[6]

Marshall readily acknowledged that his involvement in radio tracking had
been "sparked" by Warner's enthusiasm. Having spent much of the 1930s in a
variety of federal forestry and wildlife jobs before taking a job at the University
of Minnesota after the war, however, Marshall had stronger connections to the
practical concerns of wildlife managers than Warner had. Since the late 1940s
his research had focused on the population dynamics of ruffed grouse at the
university's Cloquet Forest Research Center in northern Minnesota, continuing
a study that had been begun in the early 1930s at the urging of Aldo Leopold. In
1958, he had handed over responsibility for fieldwork at Cloquet to Gordon W.
Gullion, who also had a background in fish and game management.[7]

By September 1958, the Fish and Wildlife Service office in Minneapolis had concluded that two applications of Honeywell's electronics know-how
were worth pursuing: "a tiny radio for use in studying animal migrations and
movements, and an infra-red scanning device for use in conducting large animal
censusing." Honeywell claimed that the radio tag could be built as soon as it
was "advised of the size, weight and frequency desired." R. W. Burwell of the
Fish and Wildlife office subsequently contacted James Kimball, the director
of research for the fish and game department of the Minnesota Department of
Conservation, to explore the possibility of collaboration. In October the Fish
and Wildlife Service, the Minnesota Department of Conservation, Honeywell,
and Marshall agreed to cooperate on a radio-tracking study of a ruffed grouse
population at the Cloquet Forest.[8]

Despite the apparent enthusiasm of the Fish and Wildlife Service for radio
tracking, it was clear by December 1958 that it would not provide any funding
for Marshall's project. Marshall and Gullion's ongoing grouse work at Cloquet
was funded by the Federal Aid in Wildlife Restoration Act (or Pittman-Robertson Act), through which the federal government distributed revenues from an
excise tax on sporting arms and ammunition to the states for wildlife restoration, but those relatively modest funds could not meet the cost of Honeywell's
engineering services, estimated at nearly $10,000. Nor was Marshall able to
obtain funding from the usual nongovernmental sources of support for wildlife
research. Clarence Cottam of the Welder Wildlife Foundation told Marshall
that he believed his work was of "profound importance" but that the $4,500
Marshall had requested was far above the foundation's capacity. C. R. Gutermuth

of the Wildlife Management Institute similarly told Marshall he was unable to support the project but would try to interest private donors, none of whom came forward. Marshall eventually turned to the National Science Foundation, then less than a decade old. His proposal, submitted in January 1959, described the existing methods of field biologists as "antiquated" in light of recent advances in the physical sciences. Despite nearly three decades of research using the best techniques available, research at the Cloquet Forest had failed to explain why grouse populations rose or fell. At the same time, Marshall wrote, the project's long history would facilitate the testing of electronic methods for tracking individual animals: "Field conditions are ideal in that it will be possible to work with a known group of birds on known terrain under controlled conditions using well qualified personnel."[9]

Warner also turned to NSF for funding to expand his work at Cedar Creek. In January 1959, he and Tester submitted a proposal for a study of "relationships between the movements of selected animals and some environmental factors" using radio tags. Their proposal, like Marshall's, was assigned to NSF's Environmental Biology Program, then headed by George Sprugel Jr. Sprugel was initially somewhat skeptical toward both proposals. The facilities at Cedar Creek and at the University of Minnesota seemed impressive, he told Warner and Tester, but the details of their research goals and methods were too vague. Their plan to use a "passive transponder" to track animals lacked "any details of how this will be attempted or any reference at all to similar techniques being utilized at the present time by other workers in this country." Sprugel likewise told Marshall that he needed to inform himself about similar projects elsewhere. At the same time, Sprugel was concerned about the unrestrained enthusiasm for the technique among wildlife biologists: "It seems that people are going off in all directions on the business of electronic tracking of animals." Unlike Warner and Tester, Marshall had given ample details about the technology to be used, as provided by Honeywell. Sprugel's main concern with his proposal was that the price for Honeywell's engineering services was too high. Marshall told Sprugel that he believed Honeywell's engineering expertise was worth the price and that the wildlife radio-tracking projects in progress elsewhere were taking very different approaches. Unlike Warner and Tester's passive transponders, for example—tags that would rebroadcast incoming radio signals rather than generating their own signals—Marshall intended to use active, battery-powered tags. In contrast, Warner and Tester seem to have been unable to assuage Sprugel's concerns about their proposal. At some point before the final decision was made, Warner met with Sprugel to discuss the proposal. Although one colleague of Warner's later remembered him as a charmer who "could sit down and talk you out of your shirtsleeves and underpants," the meeting did not go well. As Warner remembered it, Sprugel told him the proposal would not "stand a snowball's chance in hell in Washington," and Warner began cursing the moment Sprugel left his office.[10]

In February 1959, while NSF was still considering the proposals of the two Minnesota groups, the Office of Naval Research convened a meeting in Washington on the use of electronics in biological research, especially animal tracking. ONR's Sidney R. Galler had been promoting the development of new techniques for studying animal migration and orientation since at least 1953, when he had organized a conference on the topic, and ONR's Biology Branch had been involved in the development of radiotelemetry devices for biological research since 1956, when the Biology Branch contracted with American Electronics Laboratories of Philadelphia to build a radio thermometer. Although the thermometer was primarily intended for studying the effectiveness of cold-water suits for sailors, the biologist Carl Eklund used it to measure the incubating temperature of a penguin egg as part of the International Geophysical Year research program in Antarctica. When Galler became the head of the Biology Branch in 1957, with marine biologist Helen Hayes as his assistant, he made the development of radiotelemetry one of his top priorities. The navy also got involved in wildlife telemetry through the Naval Research Laboratory, which was developing telemetry equipment for the space program. The January 1959 issue of the *Journal of Wildlife Management* included a report titled "Design of a Miniature Radio Transmitter for Use in Animal Studies" by engineers at the Naval Research Laboratory and biologists associated with the U.S. Naval Medical Research Institute who had implanted short-range radio transmitters in woodchucks and wired their burrows to detect their movements within a 400-acre plot as part of a research program on the transmission of disease by rodents. The system's applicability to more far-ranging animals was limited, but it proved that the basic concept was viable.[11]

Galler's office was one of the central nodes in the network connecting military electronics expertise to wildlife biology. He and Hayes had invited representatives of the Fish and Wildlife Service to the February 1959 meeting along with military personnel and engineers from defense contractors such as American Electronics Laboratories. Although ONR had served as a general-purpose federal science funding agency in the years immediately after the Second World War, its mandate had shrunk with the growth of NSF in the 1950s. In his discussion of wildlife tracking, Galler emphasized the military relevance of the technique. Wildlife research had two potential benefits for the navy, according to Galler. First, it could help the navy deal with immediate biological threats, such as shark attacks, biofouling, and bird strikes. Second, knowledge gained about animals could potentially be used to improve navy technologies, including navigation and missile guidance systems. Galler and Hayes later wrote that information about bird movements obtained through radio tracking "may help discover the bird's secret of migration, which disclosure might, in turn, lead to new concepts for the development of advanced miniaturized navigation and detection systems." Galler told the assembled group that ONR was most immediately interested in a technique for studying the homing behavior of albatrosses on Midway Island, an atoll at the northwestern end of the Hawaiian Islands that was the site of a navy airfield as well as a breeding colony for several hundred thousand Laysan albatrosses. In 1957, after one aircraft sucked an albatross into its engine, the navy had been forced to temporarily suspend jet operations from the island. Galler and Hayes had previously collaborated with FWS's Branch of Wildlife Research in an effort to solve the Midway bird problem. Moving the albatrosses to a different island was one possible solution, but only if the albatrosses would stay put. Confidentially, Galler told the group that ONR had already contracted with American Electronics Laboratories to build an albatross-tracking transmitter that would initially be tested on pigeons. The FWS representatives agreed that such a technology would be especially useful for studying migratory birds and fish, and one of them mentioned that FWS was already involved in developing the technique through Marshall's grouse-tracking project.[12]

In early March 1959, Warner and Tester learned that their proposal had been rejected. Marshall was more fortunate. A few days after Warner and Tester received their rejection, Sprugel told Marshall his project had been tentatively approved, though final approval would have to wait until the next fiscal year's budget had been determined in June. Marshall was eager to begin work on a radio tag. In April 1959, Jack L. Seubert, a wildlife biologist in South Dakota's fish and game department, asked Marshall if he would be interested in cooperating in the development of the technology, which Seubert hoped to use to track deer. Marshall responded testily, asking Seubert how he had found out about what he thought was a confidential proposal; Seubert reminded him that the project had been described in a recent Fish and Wildlife Service newsletter.

After Warner's rejection by NSF, Blegen, the graduate school dean, urged him to continue seeking funding elsewhere for the use of "advanced instrumentation" at Cedar Creek, as did Spilhaus. Spilhaus sat on the board of directors of the Louis W. and Maud Hill Family Foundation, a private foundation in St. Paul, and he put Warner in touch with the foundation's executive director, with whom Warner seems to have gotten along better than he did with Sprugel. By the end of June, Marshall had received confirmation from Sprugel of an NSF grant of about $20,000, and Warner had received confirmation of a grant from the Hill Family Foundation of $40,000.[13]

These two grants placed Marshall and Warner at the forefront of a growing list of wildlife biologists interested in wildlife radiotelemetry. Kimball, the head of the Minnesota Department of Conservation's fish and game research division, began promoting both projects—especially Marshall's, in which he had a closer interest—to the press as examples of the modernization of wildlife research that would ultimately benefit the wildlife biologist's primary constituency: recreational hunters. In August 1959, the national sportsman's magazine *Outdoor Life* published an account of Marshall's grouse-tracking project and other proposed radio-tracking projects in South Dakota, Montana, and Ontario. According to the article, the ultimate goal of all these projects was "to increase the game supply and improve hunting." Kimball's promotional efforts also generated coverage of the grouse project in a number of small Minnesota newspapers and the major dailies of Minneapolis and St. Paul. The publicity was, Marshall told his partners at Honeywell somewhat disingenuously, "self generating and in fact almost ran away from me."[14]

In the fall of 1959, Marshall's team sent a mock-up of a grouse radio tag to the Honeywell office in Seattle, but it took until late October for Honeywell to assign an engineering team to the project, and it was only in November that two of Honeywell's engineers visited the field site at the Cloquet Forest, which Marshall had been requesting since March. After seeing the field conditions in person, the Honeywell team suggested giving up "the original tracking scheme of two antennae on towers," which Marshall had proposed, in favor of portable receivers carried on foot or by car, which would be more effective in locating the animals. Marshall rejected the suggestion because he thought that "carrying the latter around would cause disturbance to the birds," defeating the purpose of a system of remote observation.[15]

In early December 1959, at Marshall's invitation, Kimball gave the opening speech at the annual Midwest Wildlife Conference. Despite the establishment of various federal and state programs to promote wildlife research since the 1930s, Kimball argued, researchers had done little to make the jobs of wildlife managers easier. Only when biologists had provided useful information for managers and recreational hunters would they be justified in asking for additional funding for research. The wildlife radio-tracking projects that Marshall and Warner, working "in fields outside of the ordinary realm of most wildlife biologists," had just launched were models of how wildlife research might fulfill its promise, but such projects were all too rare. Kimball urged his audience of wildlife biologists to do their part on behalf of a nation and a culture in crisis, desperately in need of the "wholesome outdoor recreation" provided by hunting: "We must have greater knowledge and we must have it soon. Where do we find a Russian Sputnik to shatter our complacency?"[16]

If We've Got the Nuclear Bomb, Everybody Should Have It

Although Marshall had given him little encouragement, Seubert persisted in trying to increase coordination among researchers interested in radio tracking. At the March 1960 meeting of the North American Wildlife Conference in Dallas, he convened an ad hoc meeting to discuss radio-tracking techniques, which was attended by twenty to thirty people. The meeting led to the creation of a new "radio tracer branch" of the Wildlife Techniques Committee of the Wildlife Society, the main professional organization for wildlife managers. Initially the committee consisted of Marshall, Lowell Adams (a biologist affiliated with the University of California in San Francisco who edited the subcommittee's newsletter), and Frank Craighead Jr., who had just launched a study of the grizzly bear population in Yellowstone National Park in collaboration with his twin brother, John Craighead.[17]

The March 1960 meeting signaled the growing interest in the technique among wildlife biologists. The technique also began to receive increasingly widespread coverage in the popular press. As Marshall's project neared the release of its first radio-tagged grouse in April, editors from *Life* and *Sports Illustrated* and producers of television programs contacted him about sending reporters and cameramen to Cloquet. Much of the press coverage emphasized the similarities between tracking animals and tracking satellites, as in an Associated Press article published in the *St. Paul Pioneer Press* that described Marshall's radio tags as having "a radio pulse language very similar to that of space satellites." News of wildlife radio tracking also provided an opportunity for the expression of anxieties about the effect of new surveillance technologies on human social relations. In a cartoon published alongside an editorial about wildlife radio tracking in the *St. Paul Dispatch*, a transmitter-equipped male grouse confessed to a female companion, "Can't get over this feeling I'm being watched. Hope it isn't my wife!"[18]

Marshall and Gullion had told Honeywell they wanted to test the radio-tracking system in April, when male ruffed grouse at the Cloquet Forest "drummed" on logs to attract mates. During the breeding season the pugnacious males could be easily captured with mirrored traps; the birds mistook their own reflections for rival males. By early April, however, the birds had begun drumming, and Honeywell had still not delivered the transmitters or receiving system. Honeywell engineer Charles D. Canfield arrived with the equipment in mid-April, leaving just enough time to install and test the system before the season came to an end. Marshall and his team were impressed by Canfield's hard work and diligence in helping set up the system, but after he left it proved largely ineffective. The transmitters were enclosed in a rectangular metal box that fit awkwardly on the back of the grouse, and the tags weighed nearly 50 grams, or more than 10 percent of the average grouse's body weight. Gullion and Marshall's graduate students captured, radio-tagged, and released their first grouse at the end of April. After a few days, the signal suddenly disappeared, and they were unable to relocate either the grouse or the transmitter using the tower-mounted antennas, whose installation had required what Marshall described as "heroics" on Canfield's part. Ten days later they captured, radio-tagged, and released a second male grouse, and again the signal lasted only a few days. With the help of a portable receiver and antenna provided by Sidney L. Markusen, a radio engineer in the town of Cloquet, they found the grouse's body about 150 feet from the point where he had initially been released. The bird had severe bruising at the base of one wing, most likely caused when the rigid antenna caught on a low-lying branch as he attempted to fly under it.[19]

Though the failures of the system were painfully evident to researchers, they were not yet reflected in the press coverage of the project. In May Honeywell's company magazine reported that the radio-tagged grouse were "apparently unmindful of their load of instruments." By that point Gullion had already concluded on the basis of the initial tests that "we are dealing with an organism which I suspect is a pretty nicely balanced aerodynamic design—and one that

William H. Marshall holds a ruffed grouse wearing a hood and Honeywell-built radio tag at the Cloquet Forest Research Center in the spring of 1960. (Courtesy of University of Minnesota Archives, University of Minnesota–Twin Cities)

may be quite easily upset by fairly slight modifications of airflow patterns." Gullion also told Marshall that he was concerned about the effect of research on the grouse population as a whole, which had been studied particularly intensively since the late 1950s: "The fact that our populations have remained relatively stable for three years here on the forest may be the direct effect of 'too many biologists,' and have nothing to do with forestry practices or other normal population behavior."[20]

After late May, Honeywell stopped responding to the team's queries about possible improvements to the system, for reasons that Marshall would only discover months later. In June, recognizing that the tags could not be used on grouse, Marshall's team decided to test them on porcupines, which could easily carry a 50-gram transmitter package, attached with a store-bought dog harness. They also abandoned the receiving system that Canfield had designed based on Marshall's interest in monitoring radio-tagged grouse from fixed stations. Instead they used a simpler portable receiving system designed by Markusen. In June, Gullion and Robert Schwab, one of Marshall's graduate students, captured, radio-tagged, and released three female porcupines, including one mother-daughter pair. Like grouse, porcupines tended to divide their time between trees and the ground, and the researchers hoped that lessons learned from radio-tracking "porkies" could be applied to grouse. One of the most important lessons of the porcupine-tracking work was that, contrary to their initial expectations, radio signals did not necessarily travel in a straight line. With the coming of summer, the leaves on deciduous shrubs and trees had made the exact location of signals harder to pinpoint. Even with the portable locators, it was not always easy to find a radio-tagged porcupine, since signals could reflect from or refract around vegetation and landscape features. By early July, Gullion and Schwab had identified stands of white spruce and Norway pine as most likely to create misleading radio signals. The only solution was to learn the landscape and the equipment well enough to recognize "signal bounce" when it occurred and then to move around with the receiving equipment until the fixes settled down on a single accurate location—a strategy that was impossible with the fixed stations.[21]

Despite the disappointment of the initial grouse-tracking tests and ongoing difficulties with signal bounce, Marshall and his collaborators continued to promote the project in the media, cooperating with the public relations staff at the University of Minnesota to conduct a demonstration of the porcupine-tracking system in July 1960. The announcement inviting the press to the Cloquet Forest described the study as "one of the first examples of space-age techniques for wildlife research" and suggested that the technology, though developed for grouse, would "eventually be used for many other species." In August, however, when Marshall visited Honeywell headquarters in Minnesota, he learned

A graduate student demonstrates the use of the portable radio-tracking system developed by the Grousar Project in the early 1960s. (Courtesy of University of Minnesota Archives, University of Minnesota–Twin Cities)

that Canfield had been transferred to a military contract, the company having decided that wildlife radio tracking had insufficient commercial potential. Marshall later admitted that the experience with Honeywell had confirmed Sprugel's concern about the high price of its engineering services. In July he learned from Gullion that a researcher at Cornell was building radio tags for ten dollars apiece. "The problem here is that for a company so large this was a pretty small project," Marshall explained to Sprugel; "I feel we would have been better off in many ways with a smaller outfit." In the fall of 1960 Marshall left for a Fulbright fellowship year in New Zealand, and the radio-tracking project remained dormant until his return.[22]

Meanwhile, with funding from the Hill Family Foundation, Warner had established a bioelectronics laboratory in the basement of the Museum of Natural

History and begun developing a tag that could transmit the temperature and location of an unconfined cottontail rabbit back to the laboratory. For advice on electronics, Warner turned to the Heat Transfer Laboratory of Ernst Eckert, a German rocket scientist brought to the United States after the Second World War. Warner later recalled Eckert's enthusiasm when he asked him "to think now not only about the nose cones on rockets . . . and so on, but how about a rabbit, right out there? That rabbit is subject to what kind of thermal regulation?" Despite the high-powered help, Warner's group made slower progress than Marshall's. In August 1960, their first progress report consisted mainly of specifications for the kind of tag they hoped eventually to build. The development of wildlife radio tracking had been hindered, they suggested, by the fact that much "information is unavailable because of military security regulations" and, ironically, by the rapid advances in electronics, which rendered most published material obsolete. Up-to-date information could only be obtained "via personal contacts with research personnel in allied fields."[23]

Still, with the establishment of the Wildlife Society's Telemetry Committee and the release of the first progress reports from Marshall's and Warner's projects, information about radio-tracking techniques began to circulate widely. In September 1960, Eugene Dustman, an FWS staff member in Washington, told the chairman of the Wildlife Society's Techniques Committee that "those of us in the Branch of Wildlife Research are intensely interested in radio transmitting equipment which can be affixed to animals, and we shall much appreciate being placed on the mailing list to receive information coming out on this subject." The appointment of Adams and Craighead as "watchdogs" for wildlife radio tracking was, he wrote, a promising development. (Because of his move to New Zealand, Marshall was temporarily off the committee.) Earlier that month, in the name of the chief of the Branch of Wildlife Research, Daniel Leedy, Dustman had forwarded Marshall's first progress report detailing the failures of grouse-tracking efforts and the lessons learned from the porcupine tests to the directors of the Denver Wildlife Research Center, the Patuxent Wildlife Research Center, and all of the Cooperative Wildlife Research Units.[24]

As news of these early efforts spread, other wildlife biologists began to develop their own radio tags and tracking systems without the relatively large grants that Marshall's and Warner's groups had obtained. One of them was Rexford D. Lord Jr., a wildlife biologist with the Illinois Natural History Survey. Lord had been experimenting with ways of improving the accuracy of his censuses of Illinois's cottontail rabbits, with unsatisfying results. Population estimates derived from nighttime roadside censuses, for example, were characterized by "extreme variability." This was due in part to the difficulty of knowing whether any particular rabbit had already been counted and how far rabbits tended to travel during the night—the kinds of questions that radiotelemetry was designed to answer.[25]

In the fall of 1960, Lord contacted William W. Cochran, an engineer who had been collaborating with an ornithologist at INHS for several years. Since 1959 Cochran had worked in the ionospheric research laboratory headed by George Swenson at the University of Illinois, where he designed and built radio beacons for some of the first Discoverer satellites, gaining experience with small, light, robust, long-lived radio transmitters and portable radio receivers. With a budget of less than one hundred dollars from Lord's Pittman-Robertson funds and the use of equipment at the ionospheric laboratory, Cochran began building a radio tag in his spare time, aiming for the simplest, smallest transmitter possible. He later recalled that "every engineer that tackled the problem back in 1960, say, whether it was Honeywell or American [Electronics Laboratories] . . . the first thing they thought about a transmitter, they thought of a box. You start with a box, right? I mean, they didn't rack-mount it, because it was for an animal, but it was a box, and it had an antenna connector on it . . . so their mindset was wrong." Swenson later recalled that while engineers with Cochran's training and experience were common—like many other young men at the time, he had been trained in radio engineering by the navy during the Korean War—he had "exceptional skill with transistor circuits." By January 1961, he had completed a prototype with a range of half a mile; by February, he had improved the range to 1.5 miles and had begun work on a directional receiver that could be used in the field.[26]

Lord was among the fewer than thirty people on the mailing list for the first issue of the *Wildlife Telemetry Newsletter*, which was mailed in January 1961, under Adams's editorship. News about Cochran's tag spread quickly with the help of the newsletter and the annual sessions organized by the "radio tracer branch" at the North American Wildlife Conference. At the March 1961 session, Tester demonstrated the Minnesota Museum of Natural History group's prototype rabbit collar transmitter, but he also expressed frustration about the slow pace of development. The next issue of the telemetry newsletter reported his statement "that the first reaction of electronics people is that this is a very simple matter—shelf items are available right now with little or no modification. But then they reach for the items and something happens, we don't know what, that leads to a lot of electronic head scratching." Lord later recalled being embarrassed at hearing Tester and others describe the current status of their projects at the March 1961 meeting "because I knew what I had was very far superior to anyone else (thanks to Bill Cochran's genius) and it had so far cost us about $250."[27]

Despite its promising performance, Cochran's tag had yet to be tested in the field on actual wild animals. Nor was it clear that the tag would perform equally well in environments that differed significantly from the farmlands surrounding the University of Illinois. Lord recalled that Tester "arranged to send his technicians down to us to test their transmitters in the field along with ours, thinking perhaps that our soil didn't attenuate radio signals as much as the Minnesota soil." The tests confirmed that Cochran's tag outperformed the museum group's tag. When Lord radio-tagged and tracked his first rabbit on April 18, 1961, he found that the rabbit's body attenuated the range of the signal, but the performance was still impressive. In June 1961, he demonstrated the system during the annual meeting of the American Society of Mammalogists, which happened to be meeting that year at the University of Illinois, and he submitted a detailed report on the system for publication in the July issue of the *Wildlife Telemetry Newsletter*. As a result, Cochran's electronics expertise was soon in high demand. He later recalled that after the mammalogy meeting where "everybody saw Rex demonstrating that rabbit stuff . . . my phone was ringing off the hook with offers of jobs." Cochran recalled being surprised at the excitement: "To me it was all just smoke and mirrors. I mean, . . . I didn't realize it was a big deal. 'Who cares about rabbits, anyway?'" Over the summer, Cochran and Lord wrote a paper for the *Journal of Wildlife Management* and received what Lord described in one of his regular Pittman-Roberts reports as "an unusually large number of letters of inquiry concerning this system which, of course, has application to other species of wildlife."[28]

By September 1961, Cochran had built a total of ten transmitters, mostly for Lord's rabbit research. An INHS biologist named Ralph J. Ellis also used one of Cochran's tags to track a young raccoon, which proved that the system could be used for animals other than cottontails. That fall Lord and Cochran also collaborated on a study of mallard ducks with Frank Bellrose, a well-known waterfowl biologist. By this point Cochran had reduced the weight of the transmitter to 38 grams, 25 grams of which were accounted for by the battery. Although two of the three radio-tagged mallard ducks they released at a NASA satellite tracking station near campus flew only a short distance before settling down in a nearby cornfield—Bellrose had apparently neglected to feed the birds earlier that day—

one of the ducks flew back and forth over the satellite tracking station. That was useless for studying migration, Bellrose's main interest, but it gave time for Cochran, on a whim, to turn on the station's Sanborn track recorder, which recorded the variations in the radio signal caused when the duck's wingbeats and respiration deformed the antenna that was wrapped around its chest. In this instance, radio tracking was a failure, but radiotelemetry or biotelemetry—the measurement of biological parameters from a distance—was an unexpected and unprecedented success. The results were published in the prestigious journal *Science* a few months later.[29]

Cochran's simple, innovative design for wildlife radio tags and receiving equipment soon transformed the practices of both of the Minnesota groups. On August 15, 1961, after Marshall had returned from New Zealand, he met with Gullion and the graduate students working on the project at the Cloquet Forest to make plans for the coming year. Schwab suggested that what they were now calling the "Grousar" project adopt the system of fully portable, semiportable, and fixed antennas and receivers that Lord had demonstrated at the mammalogy meeting in June. Meanwhile Markusen began working on a smaller, lighter transmitter that would be suitable for grouse. Cochran's influence on Warner and Tester's group was even more immediate. They invited him to visit the group in Minnesota in the fall of 1961, and a few months later he moved to the Twin Cities to take charge of the Museum of Natural History's bioelectronics laboratory. The museum group had had some success transmitting temperature measurements from rabbits on the roof of the building to their basement laboratory with their own tag, but after Cochran arrived they shifted to his design. The group soon began field tests at the Carlos Avery Game Farm, where wildlife managers with the Minnesota Department of Conservation bred pheasants, turkeys, and other game birds, although the space-age "radionic" work of the museum group did not mesh easily with existing wildlife work at Carlos Avery. In February 1962, one of the other wildlife researchers there requested of his superior that the "Museum men" be moved out of the main laboratory building in order to free up space for other researchers and "cut down on the considerable volume of noise and commotion that goes on with the operation of their installation."[30]

Although most researchers were happy to demonstrate their radio-tracking gear to other groups, they tended to keep the designs and the equipment to themselves, especially when they had paid large sums to commercial engineers for them. Honeywell, for example, never released the designs for its grouse tag.

Cochran, however, readily shared his expertise with researchers not affiliated with the museum group. Soon after moving to Minnesota, he built a set of transmitters for Glen Sanderson, a former colleague at INHS who took the tags to Malaysia for an army-sponsored study on disease-carrying rats. (Sanderson recalled that the neck collar he initially used to attach Cochran's transmitters to the rats failed because the rat simply "shrugged his shoulders and pushed it over its head"; his wife, Beverly Sanderson, solved the problem by sewing miniature harnesses for the rats.) When they found out about his work for other researchers, Tester and Warner asked him to stop and offered him a raise. Cochran recalled that Tester "realized that this was a money-getter, having an edge in instrumentation, in getting the grant money, and he explained this to me." Tester himself recollected, "I suppose if you really look at it from a bit of a bizarre standpoint it was kind of a race to see who could really make this stuff work." Cochran later denied that pecuniary considerations had entered into his decision to build tags for other researchers; he had been convinced, he recalled, that the technology should be made available to everyone who wanted it " 'for the good of science.' . . . I was that naïve. 'If we've got the nuclear bomb, everybody should have it.' "[31]

By the spring of 1962, the radio-tracking operations of the museum group were moving into high gear, as were those of several other groups, including Marshall's. The March 1962 telemetry session at the North American Wildlife Conference was so popular that people had to be turned away for lack of space. Three groups now had functioning radio-tracking systems to demonstrate: the Craigheads with their grizzlies in Yellowstone, Marshall with his ruffed grouse at the Cloquet Forest, and Warner and Tester's group, which was testing Cochran's tag on deer at the Carlos Avery reserve. The Grousar team reported on its new harnesses and transmitters for grouse, which it had field-tested for the first time in February 1962. One of the three grouse tagged, a juvenile female, was found to have a small callus on her left knee from chafing by the harness when she was retrapped eight days after being released, but otherwise she "appeared robust and healthy and struggled in the manner of a healthy grouse." Overall, the new tags seemed to allow grouse "to continue their normal pattern of life," in striking contrast to the Honeywell tags the group had tested two years earlier.[32]

Galler had also continued to pursue wildlife radio tracking at ONR in partnership with John Busser, an engineer at American Electronics Laboratories, and to stress the relevance of the technique to military needs. The press picked up on Galler's claim that studies of animal orientation and navigation could eventually improve military technologies, as in a September 1961 *Minneapolis Tribune* article titled "Secrets of the Birds May Guide Missiles." Galler was not above playing on Cold War anxieties. In November 1961, the United Press International reported his claim that the Soviet Union was ahead of the United States in putting to use the results of studies of underwater life; the *New York Times* printed the article under the headline "Undersea Lag Is Noted," echoing then-pressing concerns about the "missile gap." Wildlife biologists and engineers who worked on wildlife radio tracking took such claims with a grain of salt. Perhaps reflecting a more general attitude toward science administrators in Washington, Cochran later dismissed Galler as "just a very nice, friendly kind of guy . . . . He was just lovable. But he never did anything, as far as I can tell, except attend the meetings and get a lot of attention because he was such a nice guy."[33]

Whatever the success of his Cold War rhetoric or the value of his overall contributions to wildlife telemetry may have been, Galler did provide substantive support to the nascent field of biotelemetry. In March 1962, ONR and the American Institute of Biological Sciences cosponsored a conference on biotelemetry at the American Museum of Natural History in New York. The conference brought together representatives of the military, industry, and academia; scientists and engineers; and laboratory and field biologists, including representatives of the leading research groups in wildlife radio tracking—the Craigheads, Marshall, Warner, Tester, and Cochran, who presented the work he had done with Lord in Illinois as well as speaking for the Minnesota Museum of Natural History group. One of the major topics of the conference was the ongoing difficulty of communication between engineers and biologists. In the conference's first substantive presentation after a keynote speech by the cybernetician Warren McCullough, Warner lauded the advances that physical scientists had made in developing instruments that were more "sensitive and comprehensive" than the "human senses" on which field biologists still mostly relied. But he also reminded his audience that the "off-the-shelf" equipment described by engineers often seemed to be located on "a most elusive shelf which, even with our combined efforts, we have been unable to find." Later in the conference, Otto Schmitt, a University of Minnesota engineer who had advised Warner in his early forays into "biotechnology" and "biophysics," warned the biologists in the audience that even the most sophisticated military and industrial telemetry equipment was unlikely to be suited to their needs: "I think you will find that much current telemetry equipment is horribly costly, clumsy, and inappropriate for biological measurement. If you can't do anything better than beg one of these telemetry systems from the military, or whoever happens to own it, go ahead and do it. But I think you will find that much smoother, slicker and biologically oriented systems are the ones that we are going to be using in the reasonably near future." Galler noted that the responsibility for failed collaborations did not lie solely with engineers; biologists were rarely willing to spend the time necessary to reach an understanding with their technical counterparts: "They would much rather pick up the phone and call the consultant, if one is available, or go to an instrument organization and say, 'Build me a black box' which may turn out to be a magnificent piece of engineering but be inadequate for the biologist."[34]

One evening during the conference Warner and Cochran came up with the idea of tracking animals by satellite. Cochran later recalled telling Warner that the radio beacon he had developed in Swenson's laboratory at the University of Illinois for one of the Discoverer satellites "was so good we can put the rabbit in orbit, and we'll hear it from down here. . . . Well, for that matter, we can put the receiver in orbit and hear a rabbit from up there." Over a few beers, Warner and Cochran sketched out what such a system would look like. As Warner later described it, a satellite-based wildlife tracking system would make possible the comprehensive, continuous, objective observation of "motile" responses of animals to their environment on a global scale. Warner submitted a proposal to the National Aeronautics and Space Administration on April 16, 1962, a month after the American Museum of Natural History meeting.[35]

While they waited for the results of NASA's review, Warner, Cochran, and Tester began working on an automatic radio-tracking system that would use tall antenna towers—much like those that Marshall's group had abandoned several years earlier—rather than satellites to detect the signals from radio-tagged wildlife. Because the conflicts between the "Museum men" and the Minnesota Department of Conservation staff at Carlos Avery had continued to worsen, they turned to Cedar Creek, where Marshall succeeded Wilcox as director on July 1, 1962. One of Marshall's first decisions as director was whether to allow the museum group to build the infrastructure necessary for the automatic radio-tracking system. The system would require the installation of two large radio towers connected to the main laboratory building by cables, as well as a significant amount of equipment and laboratory space. A 100-foot tower had already been delivered to the museum along with a navy surplus radar unit in May;

Warner hoped to buy a 75-foot tower formerly used by KUOM, the University of Minnesota radio station, with funds from the dean of the university's graduate school.[36]

In late August 1962, Warner and Cochran visited Cedar Creek with Marshall to discuss the placement of the towers, which were to be erected about half a mile apart near Cedar Creek's main laboratory building. Marshall was concerned about the towers' effect on the natural character of the site and tried to get Warner to commit to removing the them once the project was completed, but Warner argued that the towers would remain useful indefinitely, whether for future iterations of the radiotelemetry project or as observation posts to detect fires or trespassing. After talking with Warner, Marshall explained to the associate dean of the University of Minnesota's Institute of Agriculture that he had "made it clear that I think this is a very fine project and will do everything possible to expedite it—keeping in mind my responsibilities to Cedar Creek as a 'Natural Area.'" When the office came through with funding later that fall, Marshall again urged Cochran to ensure that the towers for the automatic radio-tracking system at Cedar Creek were "erected with the least possible disturbance of the habitat."[37]

In October, while still waiting to hear from the dean's office about whether it would fund the installation of the towers, Warner and Cochran drove west from the Twin Cities to the Sand Lake National Wildlife Refuge in South Dakota. Cochran and Dennis Raveling, a graduate student of Warner's, had cut a hole in the roof of their university-owned station wagon so that a tracking antenna could be rotated from within the car, as Warner later recalled. At Sand Lake, with the help of refuge staff, they fitted a number of blue snow geese with radio tags that they hoped would allow them to track the birds as they migrated south across the Great Plains toward Mexico. Some twenty-five years later, writing to Warner on the occasion of his retirement, Raveling enthusiastically recalled "assisting you and Bill Cochran while listening to radio Moscow and radio Havana in between monitoring radio-instrumented geese in the first successful field test of telemetry of wild, long-distance migrants at Sand Lake, South Dakota during the Cuban missile crisis in October 1962!" Warner recalled that Cochran had been "petrified by the idea the Russians were going to annihilate us through Cuba" and returned home to his wife in St. Paul in advance of the rest of the researchers. The study was in fact less of a success than Raveling remembered. Despite the impending winter, the geese refused to migrate, and the researchers returned to Minnesota after several weeks with little to show for it except for proof that the geese would wear the harnesses without notice-able changes in behavior. Tracking the birds by satellite would have allowed the researchers to return to Minneapolis as soon as the tags were attached and still acquire migration data without the trouble of a high-speed chase to the Mexican border. Later tests on high-atmosphere balloons by Cochran suggested that transmitters and power sources small enough to be carried by birds would still be powerful enough to be detected from space. When Warner mentioned the possibility of wildlife satellite tracking to a meeting of the Minnesota Orni-thologists Union later that fall, the audience was, according to a later report, "startled," as perhaps were NASA's reviewers, who had by then rejected War-ner's grant proposal.[38]

In addition to the question of the towers, the museum group's interest in following radio-tagged animals wherever they might go brought them into con-flict with Marshall's interest in protecting Cedar Creek as a "Natural Area." In early 1963, while the automatic system was still under construction, Marshall questioned Warner's claim "that the motile response project needed access to the entire Natural History area at all times." Marshall reminded Warner that he had originally agreed that the group would not operate in the Cedar Bog Lake area, the core of the reserve. In his reply, Warner justified the radiotelemetry project's request for access to "the area that is now designated as inviolate by man" in terms of the need to observe radio-tagged animals in the flesh; mere re-mote observation would not suffice: "If, for example, signals from a radio tagged animal indicate that an animal is motionless in the inviolate area, and remains so for an *unusual* length of time, we should like to be able to walk to the site in order to determine the reason for that behavior. . . . It is my sincere belief that the field biologists and others on this project recognize the need to preserve wild areas in as near their natural state as possible." Warner later recalled that Cedar Creek was "the logical place" for the museum project, despite the need to "battle" restrictions on handling animals and using the entire landscape: "Even-tually we convinced even old Wilcox and so on we weren't going to massacre the habitat."[39]

Despite advances in the use of the technique, Marshall's early failures to track grouse in the spring of 1960 had convinced some sportsmen that the technique was a waste of money. In July 1962, the head of the Minnesota Emergency Conservation Committee, Olin Kaupanger, wrote Kimball to express his doubts about all academic research into wildlife, particularly "radio-sending devices." A column in the *Minneapolis Star Tribune* that winter similarly questioned the value of wildlife research. As Marshall quoted it in a letter of protest to the newspaper's managing editor, the article had argued that "grouse can't be helped under any known program (the biologists' radio experiment to track grouse failed) so not much can be done there." The editor agreed to print an article by Kimball defending the project under the headline "Electronics Permits Intimate Insights into Animal Life," which included a photo of Marshall with the most recent version of the radio tag. "Infinitely more important than this particular study is the development of radio-telemetry, a technique which can lead to better understanding the intricate lives of wild animals," Kimball wrote. "Progress is based on knowledge, whether it be in the field of rocketry or wildlife management."[40]

Despite the skepticism of Kaupanger and the *Star Tribune* columnist, interest in radio tracking among wildlife biologists continued to grow. Having been forced to turn people away from the March 1962 meeting for lack of space, the organizers of the March 1963 telemetry session scheduled a room with a capacity of 250 people, which "bulged" with the approximately 300 who showed up to hear Marshall, Tester, Warner, Cochran, the Craigheads, and several others present their latest technical advances. At the session, Vincent Schultz of the Atomic Energy Commission encouraged wildlife biologists to apply to the AEC for funding. As the telemetry newsletter reported it, Schultz told the crowd that the "AEC has money available for research grants, particularly studies involving ionizing radiation, many of which can be fitted into studies of naturally ranging wildlife species. He pointed out that there was money for both gadgets and research if the studies were framed in the proper way." After the conference, Tester submitted a proposal to the AEC to use the automatic radio-tracking system at Cedar Creek, then still under construction, to study the effects of ionizing radiation on wild animals. Tester later recalled that the AEC "didn't really insist on or even care especially about using radiation as an environmental stress. They liked the fact that we were sort of pioneering in the telemetry area and . . . their thought was that this is going to have benefit to ecological studies."[41]

Tester's ultimately successful application to the AEC marked a significant shift in the museum group's work, which until then had been led by Warner and funded primarily by the Hill Family Foundation. Though Warner was more senior than Tester and eventually succeeded in obtaining a large grant of his own from the National Institutes of Health to provide stipends for graduate students working at Cedar Creek, personal health problems rendered him incapable of leading the group, and Tester filled the vacuum. Looking back four decades later, Warner described Tester as an aggressive "go-getter" who "really jarred a lot of people" and as someone who "can get a proposal off on your ideas in about four minutes," but he was grateful to Tester for keeping the project going when he was incapacitated: "As I look back, and I've told John this, 'John, if it weren't for you, we might not have kept it all going.'" Tester himself recalled that his main contribution to the project lay in his understanding of grantsmanship: "I guess my expertise was raising money." From 1959 to 1965, the Hill Family Foundation provided a total of $166,134 for Warner and Tester's wildlife radiotelemetry work. That sum was soon dwarfed by the grants from the AEC and NIH.[42]

Nothing More Than Natural History?

As wildlife radio-tracking designs and equipment became more widely available, even some of the most ardent advocates of the technique began to express doubts about its scientific value. In the December 1963 issue of the telemetry newsletter, Adams described two forms of "disillusionment" with the technology. Radio-tracking equipment was "not so easily acquired as expected," and "the technology has given no evidence that it is anything more than a curiosity as a biological research tool." Now that the initial excitement had worn off, he suggested, wildlife telemetry was entering its "shakedown period." Adams expressed similar concerns privately to John Olive of the American Institute of Biological Sciences, who had suggested creating a new venue for publishing research in biotelemetry. In Adams's opinion, there were already ample venues for disseminating reports on radiotelemetry, most of which remained focused on gadgetry; what was needed now were "more reports of more scientific results." Two groups seemed closest to producing such results: the Craighead brothers in Yellowstone and Marshall's team at Cloquet.[43]

The Grousar project team had spent the first three years of the project developing tags that would not significantly affect the behavior of grouse and a receiving system that would allow them to be tracked effectively. Although they had radio-tracked a handful of male grouse for about a month in April 1962, their first season of extensive tracking of both male and female grouse was not until 1963. By then Markusen had managed to reduce the weight of the transmitter, and Gullion and Marshall's graduate students had significantly improved the harnesses used to attach the transmitters to the grouse. Despite these improvements, the Grousar progress report for 1963 noted that, of the fifteen birds tagged that year, three "exhibited an abnormal behavior which was quite probably due to an aversion for the packages they carried." For example, Number 1817, a juvenile female, moved erratically after being tagged on March 4, 1963, refusing to flush from a dense stand of balsam fir, where she remained for three days, and then returning to the area where she had been trapped. By March 22, when she was recaptured and her transmitter removed, she had lost more than 15 percent of her original weight. The Grousar team concluded that "#1817 is a bird who is behaviorally unsuited for carrying a radio." Dividing the population into birds who were "behaviorally unsuited" and birds whose behavior remained "normal" allowed the team to avoid—though only temporarily—the question of whether grouse were being affected by the tags in more subtle ways.[44]

In the winter of 1963–1964, Marshall's team ran into an unexpected hurdle that made it impossible to achieve the scientific results Adams was hoping for. The population of grouse at Cloquet plummeted, and Gullion warned against attempting to trap any of the remaining grouse, who were at risk of injuring themselves or being trampled by the snowshoe hares that sometimes also entered the traps. Marshall instructed one of his graduate students to radio-tag the hares instead, but the results were disappointing; the hares covered very small ranges and were most often discovered hiding under windfalls. The radio tagging of grouse resumed in the spring and summer of 1964, when the population rebounded. Another of Marshall's graduate students, Geoffrey Godfrey, was interested in studying the breakup and dispersal of broods in the summer and fall. To do so, he radio-tagged young birds and tracked them by car and on foot, extending the study into private lands bordering the Cloquet Forest. By the end of 1964, the Grousar team had radio-tagged thirty-five different grouse, some of them multiple times, since the beginning of the project.[45]

Though it had initially lagged far behind Marshall's group in deploying radio tags on animals in the field, the museum group led by Warner, Tester, and Cochran quickly leaped ahead with the help of an influx of funding from the AEC and NIH. Cochran completed Cedar Creek's automatic radio-tracking system in 1963 with the assistance of Larry Kuechle, an engineering undergraduate at the University of Minnesota who would later take Cochran's place as head of the bioelectronics lab. The 75-foot tower became operational in July 1963, and the 100-foot tower in November, along with all the necessary receiving equipment at the laboratory, much of it military surplus. The museum group also expanded its biological staff. Alan Sargeant, an FWS biologist in North Dakota, had been deputed to the project in July 1963, and L. David Mech, who had studied wolves as a graduate student under wildlife biologist Durward Allen at Purdue University, joined full-time in September. On October 24, 1963, Olive convened a meeting on wildlife telemetry at the American Museum of Natural History, which was attended by both Warner and Gullion. After returning to Minnesota, Gullion told Marshall that the "Cloquet project is still the most successful in the country, but we're not getting the word out and selling it the way some other projects are, notably the Minneapolis Museum project."[46]

Unlike Marshall's team, which was interested in using radiotelemetry to understand the life history and population dynamics of ruffed grouse, the Cedar Creek group focused on developing the technology. Tester later recalled that "we were trying to work with anything that was big enough to carry a radio transmitter at Cedar Creek." Except for studies of birds and fish, Tester recalled, most of the attachments were easy: "A collar around the neck of an animal is like putting a collar on a dog or like putting a watch on your arm." Mech later noted, "It's not that we set out to find out a specific thing. Rather we said, 'Here's this major new revolutionary technique; let's see what we can get out of it.' . . . We were exploring all around trying to do this and that just to see what was possible to do. It's the opposite way that you would ordinarily design a study, but given that this was a revolutionary technique it made perfect sense to do that." Cochran, the engineer, suggested that this approach was far from unusual: "I think, in general, science—it's the old adage, which the head and which the tail. The history of science often follows the instrumentation." The first animals tracked with the automatic system were deer, red fox, cottontail rabbit, and badger. All of these species were able to tolerate larger, heavier, less carefully attached tags than Marshall's grouse were, and all of them were easier to trap, handle, and track than the Craigheads' dangerous, far-ranging grizzly bears were. Still, even easily handled animals capable of carrying large tags could present unexpected difficulties. As another researcher who began radio-tracking raccoons around this time noted, his subjects were often "exasperatingly clever at divesting themselves of collars and harnesses." In the early days of radio tagging at Cedar Creek, Mech recalled, a major challenge was keeping the collars on the animals: "Every now and then we'd get one back—you know, you'd have to live-trap them and everything—and we'd find out that they were scratching through the collars." Mech, who played an important role in getting the automatic system working as an actual research instrument after Cochran and Kuechle finished building it, claimed that his most significant contribution to the project was the use of dental acrylic to make the collars more durable.[47]

The level of concern about the effect of tagging on animals varied among researchers, as did the reasons for concern. Most were concerned mainly with preventing an ill-designed tag from biasing their data. Warner recalled telling his students who were interested in radio tracking: "You don't have to perfect it, but make sure you've thought about what kind of a shape an animal is, and how it moves, you know, flying, walking, and so on. Be sure you take all those into account when you build a harness, or whatever you build." Cochran was especially concerned about the effect of radio tags on the migrating birds he was most interested in studying. He later described himself as "squeamish," in contrast to many of the biologists he had worked with. He never liked "to put transmitters on birds. It's traumatic. You see yourself interfering with his life. Then he's gone. You don't think of him as a living being. He takes on a different place in your mind. He becomes a data machine—where he is, how fast he's going." Cochran's first mentor and collaborator in ornithology in Illinois, Richard Graber, expressed similar concerns about trapping and tagging birds in his own writings. But few of the biologists involved in the early development of radio tracking could be described as squeamish. Most of them had grown up trapping, fishing, hunting, or working with livestock. Tester described himself as growing up as an "avid hunter," and Warner hunted regularly and attributed his interest in animals' use of energy to growing up on a dairy farm in frigid Minnesota. Marshall had spent some of his spare time during his Fulbright year in New Zealand hunting big game such as chamois and red deer. For people used to handling and killing animals for work and for pleasure, tagging could easily seem like an insignificant intervention.[48]

Concerns about the distortive effect of the local environment also varied widely among researchers. In contrast to the Grousar project, which had struggled to account for signals reflecting off or refracting around stands of trees and other landscape features, the Cedar Creek team experienced relatively little trouble with signal bounce. At the October 1963 telemetry meeting in New York, while its system was still under construction, Warner had claimed that the museum group experienced no such problems at all. In May 1964, after it had been operational for about six months, Cochran, Warner, and Tester reported that they had experienced none of the problems that had plagued the early work of the Grousar project. Kuechle later recalled that "one of the reasons that we

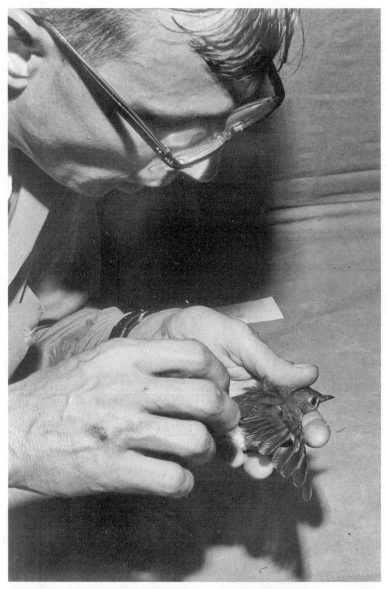

After returning to Illinois from Minnesota in 1964, William Cochran began studies of the migration of thrushes, which required extremely miniaturized radio tags. (Courtesy of the Illinois Natural History Survey)

didn't have, at least early on, much trouble with it at Cedar Creek was because we had those tremendously high towers." Two years after the system's launch, Sargeant, the FWS biologist who was using it to track foxes, reported that some signal distortion did occur, but he also concluded that its overall impact was "negligible," at least given the movement patterns of foxes. Kuechle later noted that

> most of the time you can mitigate it because once you've developed some track-ing experience you know when it's happening, because . . . you have maybe some signal coming from the direct path, and some signal coming from some other path, and so when you try to find a bearing on it, the bearing is not—it just doesn't seem right, you know, it just doesn't have the right feel. And with experience you know that it's happening, and so the solution is to simply move someplace else and get away from that obstruction, which you can do fairly well in Minnesota, but it's more difficult in a mountainous area.

Still, the Cedar Creek system was subject to other kinds of problems that com-plicated the interpretation of radio signals. Damaged power lines in the area around Cedar Creek occasionally produced electrical interference, requiring a call to the local power company, and on windy days, the accuracy of the readings was reduced by the sway of the towers.[49]

Even researchers with extensive experience with radio tracking could be surprised when they shifted to new field sites or new species. After Cochran returned to the Illinois Natural History Survey in May 1964, he designed an automatic radio-tracking system in southern Illinois that was modeled closely on the Cedar Creek system. Two INHS biologists tried to use it to track trans-located deer, but the hilly, heavily forested landscape—very different from the flat, sparsely forested landscape of Cedar Creek—limited its accuracy and effec-tiveness, and it ultimately received little use. In 1965, Cochran collaborated with Graber on a radio-tracking study of the migration of thrushes, which weighed 30–40 grams, perhaps a tenth of the weight of the average ruffed grouse. For these birds, Cochran designed an ultraminiaturized radio tag weighing only 2.5 grams, which was still a significant fraction of the birds' weight. In the conclud-ing paragraph of the paper reporting on the study, Cochran and his coauthors explained that one ornithologist, seeing that the radio-tracking findings con-tradicted some of his own observations, had "suggested that the radio fastened to the birds was responsible for some of the behavior we observed. This point is

In a paper on thrush migration published in 1967, Cochran and his colleagues reminded their readers that "studies using radio-tags are studies, not of birds, but of radio-tagged birds." An antenna can be seen protruding from under this bird's feathers. (Courtesy of the Illinois Natural History Survey)

well taken, for the near impossibility of providing controls dictates that studies using radio-tags are studies, not of birds, but of radio-tagged birds."[50]

Researchers also struggled to manage the massive amounts of data that radio tracking could produce. The Cedar Creek system was designed to collect location data from dozens of animals at fifteen-minute intervals. Cochran, Warner, and Tester argued that automation was necessary because "several persons could not continuously monitor the position of even one animal under natural conditions" using portable receivers, much less the "exact location at all times" of multiple animals. The system that Cochran designed featured rotating antenna arrays on top of each of the two towers, which were connected by electrical cables to the main Cedar Creek laboratory building, where they were fed into a bank of receivers, each tuned to a different frequency corresponding to one of the tagged animals. A continuously running 16-mm film recorded the changing signal strength displayed on the receivers. After the film was developed, the direction of peak signal strength could be read from the film for each of the two towers and used to triangulate the location of the animal. In 1964, the system was used to track a total of sixty-eight animals: forty red foxes, three grey foxes, six white-tailed deer, eleven snowshoe hares, and eight cottontail rabbits.[51]

The automated system changed the nature of the labor required to collect and process data rather than eliminating it entirely. The developed film was usually transcribed into map coordinates by Beverly Bonde, who also performed secretarial duties at Cedar Creek headquarters; the author of an account of the history of Cedar Creek written in the 1980s thanked her for "tolerating so faithfully the tedium involved in locating the positions of animals from film on which the signals were recorded." Bonde sometimes shared the tasks with other "film reading girls" and with the researchers themselves. Soon after the system had become operational, Cochran determined that it took him nine hours to process two days' worth of the movements of a single fox. A later report claimed that twenty-four hours' worth of movements could be plotted in six hours—three hours to transcribe the locations from film and three hours to plot them on a map—and concluded that, although the system was functional, the "need for some automation in data processing is obvious." To simplify the "reading" process, Donald Siniff, a statistician who was hired under Tester's AEC grant, later recalled creating a data sheet for film readers that included the identification number of the animal, date and time, x and y coordinates, and whether or not the animal was active, which could be determined from small variations in signal strength. Other kinds of nonautomated labor included developing and reloading film and tuning the receiving system to account for drift in the frequencies of transmitters.[52]

The need for someone with Siniff's statistical skills was recognized early on by the designers of the Cedar Creek system and by their federal sponsors. As Siniff later recalled, the AEC's Schultz had told Tester that he would only fund the project if the museum group had a statistician on staff. Siniff had met Schultz earlier through Lee L. Eberhardt, a statistician for whom Siniff had worked at the Michigan Department of Conservation. Six months after the meeting, according to Siniff's recollection, he received a terse postcard: "Tester at Minnesota needs a statistician—Schultz." Siniff applied for the job and began working at Cedar Creek early in 1964 while enrolled as a Ph.D. student at the University of Minnesota. After meeting with Tester at the North American Wildlife Conference in Las Vegas a few months after Siniff's arrival in Minnesota, Eberhardt suggested to Siniff that "computer technology" might be the best way to "handle the remarkably large number of observations which can be easily generated" with an automatic radio-tracking system. In October 1964, Sargeant gave Siniff a list of parameters that he wanted to have calculated automatically from the raw location data of the foxes he was tracking. Siniff began working on FORTRAN programs to calculate the distance traveled by an animal, its center of activity, the mean distance from a given reference point, and the intensity of an animals' use of various parts of the landscape. He also developed a system for coding the data read off the films onto punch cards that could be fed directly into the university's mainframe computers.[53]

Siniff's work on computer programs for analyzing radio-tracking data led him to experiment with simulations of animal movements, which became the focus of his doctoral thesis. Through his mentor Eberhardt, Siniff was connected to the system of national laboratories that were one of the legacies of the Manhattan Project, where ecosystems ecology and systems analysis were dominant. But these theoretical frameworks had little influence on the way Siniff and other biologists at Cedar Creek thought about their research on wildlife. After finishing his doctoral thesis in 1967, Siniff told Eberhardt, "I suspect if one tried, you could put what I tried in my thesis in terms of systems analysis, but [I] really see no need as long as you can think in terms of a series [of] events. I take this to be considerably different than the types of things they fiddle with at Oak Ridge." Instead, Siniff, like other wildlife biologists, grounded his work in the idea of "home range," a term coined by naturalist Ernest Thompson Seton in the early

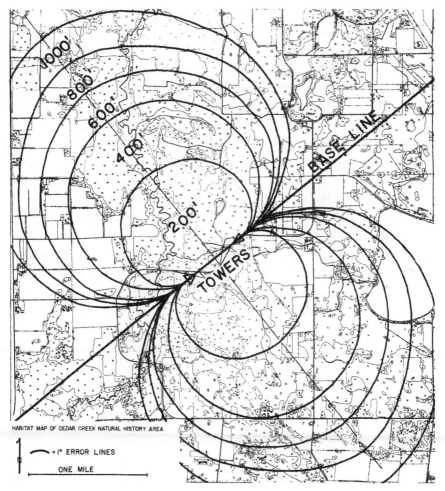

HABITAT MAP OF CEDAR CREEK NATURAL HISTORY AREA

= 1° ERROR LINES

ONE MILE

The location of the two antenna towers for the automatic radio-tracking system at the Cedar Creek Natural History Area created areas in the landscape where animals could be tracked with greater or lesser accuracy. (Source: Alan B. Sargeant, James E. Forbes, and Dwain W. Warner, *Accuracy of Data Obtained Through the Cedar Creek Automatic Tracking System*, Minnesota Museum of Natural History, Technical Report Number 10, December 1965, 20 pp., on p. 12, courtesy of the Bell Museum of Natural History)

twentieth century and redefined in quantitative terms by mammalogists such as Lee Dice and Don Hayne in the 1940s.[54]

Studies of home range aimed to quantify and map an animal's regular or habitual movements so that they could be easily compared to those of other animals. Until the development of radio tracking, home ranges had most often been mapped on the basis of trapping data, which usually provided only a handful of data points for any particular animal. Radiotelemetry, especially using an automatic system such as Cedar Creek's, massively expanded the number of "fixes" or locations of an animal's movements. As the resolution of data about animal movements increased, however, so did skepticism about the very concept of home range. In 1966, the *Journal of Wildlife Management* published a review of techniques for studying mammal movements by Sanderson, the INHS biologist who had taken some of Cochran's transmitters to Malaysia for a study of disease transmission by rats several years earlier. Sanderson noted that radiotelemetry was "one of the more exciting recent techniques for studying mammal movements" even though it had as yet "made few contributions to the fundamental problems confronting students of mammalian ecology." But regardless of the technique used, Sanderson was skeptical of abstract measures of the size or shape of home range that were divorced from the natural history of the animal and its habitat. To understand a cornfield as a habitat for cottontail rabbits, he argued, one needed to know "whether the cornfield is next to timber, mowed or unmowed hay, soybeans, or a roadside, the time and weather factors, the cornfield's own special qualities, and so on"—not just the size or shape of the animal's home range.[55]

The Cedar Creek group was well aware of the limitations of the home-range concept, even as they relied on it to simplify their analyses. One paper by Tester and several colleagues noted that "within a home range there are usually several areas of intensive utilization or centers of activity; these may shift in time, depending on individual or environmental variables." Siniff later recalled that he was never satisfied with the definitions of home range he found in the literature: "All these behaviors that are incorporated into what you might call home range are just behaviors that life histories have, and you just never quite can generalize them to some kind of a concept." Cochran, too, recalled that over time "the term itself became questionable."[56]

Siniff's simulation work was based on the insight that each "fix" in a series of animal movements could be understood as a point of decision for the animal about how far to move and in which direction. Statistical distributions of angles and distances of movement derived from actual animal movement data could be used to generate "decisions" at random, and the resulting simulated maps could then be compared with real movement data. Eberhardt encouraged Siniff's simulation work, though he was skeptical of the level of abstraction it demanded, arguing that "we likely cannot expect any simple distribution to exactly represent the behavior of an animal, but it does seem that there ought to be something there." Siniff's hope was that even oversimplified models would provide insight into how animals used space and that deviations in real animal data from the simulations would indicate "factors influencing the animals' decisions." Siniff later recalled joking with researchers when the automatic radio-tracking system had gone down, producing gaps in the movement data for individual animals: "I'd say, 'Don't worry. You want fox data? I'll make you fox data. You want grouse data, I'll make you grouse data.'"[57]

Even as field techniques and data analysis procedures became increasingly sophisticated, skepticism about the scientific value of the technique continued to grow. Warner later recalled being frustrated that his students were only interested in "following animals around," rather than pursuing the kinds of questions about energy use and microclimates that had initially gotten him interested in telemetry. Cochran shared Warner's concern, even as he recognized that Warner's own approach to research was often driven more by a desire for more data than by any clear idea of what to do with that data once he had it. In January 1965, as guest editor of the *Wildlife Telemetry Newsletter*, Cochran expressed his hope that "as the field biologist tires of movement studies and the lab biologist starts wondering what his subject would measure on the loose, the know-how of each area will merge at its own pace."[58]

By the mid-1960s, federal funding agencies had also become leery of projects that seemed focused on gadgetry rather than on important scientific questions. When Marshall submitted an application to renew the Grousar project's grant to NSF in January 1965, the Environmental Biology Program's new director, John Rankin, told him it would need extensive revision. Sprugel, who had been a consistent if critical supporter of early radio-tracking projects, had left NSF the previous fall to become the National Park Service's chief scientist. The panel that reviewed Marshall's revised proposal unanimously criticized the Grousar project for its paucity of published scientific findings and its use of an expensive, complex technology where cheaper, simpler ones would have sufficed—perhaps binoculars, one reviewer suggested. One reviewer accused Marshall and his team of not really conducting modern science at all: "This is nothing more than natural history." Another wondered why the Fish and Wildlife Service could not take responsibility for funding the project. All three reviewers recommended against renewal. In March 1965, Marshall visited Washington to discuss the project with Rankin and the Environmental Biology Program's long-standing assistant, Josephine Doherty. In a follow-up letter, Marshall defended the scientific value of the project, pointing to a recent article he had authored for the journal *BioScience* that showed "very clearly that the radio-telemetry technique is far advanced over . . . older methods." It was difficult, he added, to get funding from state fish and game departments for "research as intensive and basic as this." Despite Marshall's efforts, official notification of rejection arrived in April, and the Grousar project's last day of fieldwork took place in August.[59]

Although Marshall was eager to apply for a new grant to sustain the project, Gullion strongly discouraged him from doing so. Too much data had been collected but not analyzed, he argued, and no one really knew what had or had not been accomplished. Moreover, Gullion had begun to doubt the validity of much of the data collected by radio tagging. Many male grouse seemed to drum less frequently after being tagged. After seeing a proposal for a new grouse-tracking project by one of Marshall's graduate students in early October, Gullion told Marshall that the key question was

> just how much experimenting can be done with birds in a population where we are attempting to measure the response of the total population to changes in the forest environment? Normal trapping and banding raises the loss rate appreciably, and the additional handling and retrapping attendant to the radio technique increases this even more, then there is the problem that a significant proportion of the birds (mostly males) will not carry the radios. It boils down to simply a matter of whether we want to learn a good deal about grouse life history at the expense of evaluating their overall responses to forest management, or is the effect of forest management on the population our major concern?[60]

Without the support of Gullion, who supervised the day-to-day grouse research at Cloquet, it would have been extremely difficult to continue the project. Marshall began working on his final report to NSF and urging his students to turn the data that were still "locked up in thesis" into publishable articles. In his final report, Marshall did not emphasize the values of objectivity and space-age, Cold War–era modernity that had appeared in his initial grant proposal in 1959 or that had characterized much of the subsequent press coverage. Instead he claimed that the "system has one basic attribute. The biologist becomes, in a very real sense, intimately acquainted with the animal carrying the transmitter and also with the habitat it is occupying, as the work proceeds. Thus, he can do a great deal of qualitative interpretation on the spot adding to his understanding of the complex conditions encountered in the field." Such "qualitative interpretation" was useful less for gaining a holistic view, Marshall explained, than as a way to "attack a specific phase in life history or one aspect of the ecological and behavioral problems involved." Although Marshall reported that radio tracking had produced new scientific discoveries about the winter movements, breeding behavior, nesting behavior, and brood breakup and dispersal of ruffed grouse, he mostly emphasized the project's technical accomplishments, noting that the equipment it had developed was being used by at least fifteen other research groups in the United States and Canada and that several hundred copies of each progress report had been sent to researchers in the United States and abroad. Although Marshall continued to dabble in radio tracking until the mid-1970s, the end of NSF funding for the Grousar project effectively brought to a close his role as a leader in the development of the technique.[61]

The Cedar Creek group, though far more successful, also faced skepticism about the scientific value of radio tracking. Some of the other scientists who used Cedar Creek believed that radio tracking had an unacceptably high impact on its animal populations and habitats, which were intended to be preserved as examples of untouched nature for the purpose of observational research and as baselines against which landscapes and populations that had been manipulated by humans could be compared. Such concerns came to a head in 1965. Tester had justified the AEC's funding of the Cedar Creek automatic radio-tracking system on the grounds that the system could be used to study the effects of sublethal ionizing radiation on the behavior of unconfined wild animals. By late 1965, he was ready to begin capturing snowshoe hares and raccoons at Cedar Creek, exposing them to the cesium-137 source at the University of Minnesota's Gamma Irradiation Facility, and releasing them back at Cedar Creek. Writing to Marshall for permission to carry out the irradiation, Tester explained that almost nothing was known about the behavioral effects of irradiation outside of the laboratory, even though they had the potential to be devastating. If irradiation impaired courtship behavior, for example, "the overall effect of this radiation on the population would be equivalent to that of sterility or death." But Tester was also careful to note that the potential effects of the study were on the same scale as various natural factors. Irradiating a handful of raccoons and hares would, he wrote, "not be any more detrimental to the populations at the CCNHA than a normal occurrence of disease, parasitism or injury due to natural hazards."[62]

Recognizing that the study represented an unprecedented intervention into the Cedar Creek landscape, Marshall asked several University of Minnesota biologists to review Tester's proposal. Would enough animals be irradiated to affect the health of Cedar Creek's animal populations? Was there any risk of introducing mutations into the gene pool? One of the reviewers, botanist Donald B. Lawrence, had, like Wilcox, played an important role in the establishment of Cedar Creek; in 1950, he had personally donated 130 acres for inclusion in the reserve. Lawrence's position on experimentation at Cedar Creek was complex. A. C. Hodson, who was also involved in Cedar Creek's founding and who later compiled a history of the site, noted that Lawrence had "voiced a strong objection" to a study that involved collecting mice at the site in the 1950s. Decades later, Lawrence continued to argue that the site's main function was as a nature preserve for observational research and that "irresponsible researchers" were as grave a threat to the integrity of the site as "outside interference." In 1957, however, Lawrence himself had sought and received approval for Cedar Creek's first experimental study, arguing that that observational research alone could not address all of the important questions in field biology.[63]

Either because of or despite his own role in introducing experimental methods to Cedar Creek, Lawrence proved to be the irradiation study's fiercest opponent. Tester's proposal was, he argued, "contrary to the major objectives of a nature preserve, which are to preserve natural habitats and populations as norms or standards for comparison with habitats and populations of adjacent areas which have been artificially influenced in various ways by man." All such areas were, he admitted, somewhat "contaminated" by influences from outside the preserve, but one should make one's best effort to minimize such contamination. The national laboratories at Brookhaven, Oak Ridge, and Argonne would be more appropriate than Cedar Creek for the work Tester had proposed, he suggested. Against Tester's vision of Cedar Creek as a site for simulating a nuclear future, Lawrence posed his own vision of the site as a threatened remnant of a natural past: "To purposely manipulate the genetic makeup of individuals by radioactive or chemical or other means and to return them to the freely roving population of the Natural History Area seems to me to be initiating the deterioration of a priceless natural resource."[64]

In his response to the questions and criticisms of Lawrence and the other reviewers, Tester noted that no more than six raccoons and six snowshoe hares would be irradiated over the course of the three-year study—an insignificant fraction, he argued, of the total number at Cedar Creek. The issue of contaminating the gene pool was trickier, he admitted. Since courtship behavior and breeding were central to the research questions, irradiated animals would have to be present in the wild population during the breeding season. Gonads could not be shielded from irradiation, as one reviewer had suggested, since "we are attempting to simulate conditions which might occur in the wild in the event of an atomic war or as the result of an accident in a nuclear reactor." Even after seeing Tester's response, Lawrence remained skeptical. If animals were to be irradiated, he urged that they be females, so that any offspring could be easily identified and removed from the population. He was also critical of the experimental design of the study, in which each irradiated animal was to be paired with a sham-irradiated animal. Behavior patterns differed too widely among animals in "wild genetically heterogeneous populations," he argued, for any such pairing to be valid.[65]

The status of the irradiation study remained uncertain until the end of March, when a newly reconstituted Cedar Creek Advisory Committee met to discuss the proposal. Lawrence reiterated his belief that "the preservation of CCNHA as a living museum must be the prime consideration," and the chairman of the committee, Alan J. Brook, agreed that "experimentation should never be allowed to interfere or destroy this concept." Unwilling to jeopardize Tester's AEC grant, the committee decided to approve the study with several qualifications: any progeny of the irradiated animals would have to be sterilized or removed from the population, the researchers would have to prove that they could successfully tag and retrieve raccoons before any irradiated hares were released, and the researchers would have to consult with the director, Marshall, about any changes in procedure. The committee also agreed that future research proposals should be sent to the director of Cedar Creek for approval before being submitted to funding agencies. This latter provision was, in effect, an admission of the committee's limited authority. Once a project had been funded, it was extraordinarily difficult to block it on the basis of the damage it might cause to a natural reserve.[66]

Even as the irradiation study got under way, the museum group was becoming aware of problems with the automatic system that were inherent to its design and ultimately limited the value of the data it could collect. In particular, since the potential error in the triangulated location of a radio-tagged animal depended on where it was in relation to the towers, there were systematic biases in location accuracy across the landscape. The "error polygon" around the actual location of the animal was smallest when the lines from the towers to the animals intersected at right angles, which created an hourglass-shaped area of acceptable accuracy around the towers, as Tester and graduate student Keith L. Heezen showed in 1967. As Mech later recalled, errors outside of those areas "got worse and worse, and the farther you got out the worse it got," which was especially problematic for tracking animals such as foxes or deer that "were ranging far and wide." Tester and Heezen concluded that the automatic system "allows unbiased sampling of movements in relation to vegetation, roads, and other deer, but not in relation to the triangulation stations," although they also noted that their analysis of error was irrelevant when, as was often the case, "the purpose of the radio location is to allow the investigator to come into visual range of the animal to make observations on movement and behavior."[67]

Around this time the Wildlife Society's Telemetry Committee came to an end. The level of attendance at the North American Wildlife Conference telemetry sessions had been ebbing since 1963, as had contributions to the telemetry newsletter, even though the number of subscriptions to the free publication had risen to nearly six hundred. In the December 1966 issue, Marshall editorialized that "it is time for workers using radio-telemetry techniques to present data which will stand on scientific merit in recognized society meetings and/or publications." The Wildlife Telemetry Committee was officially disbanded in September 1967, when the last issue of the newsletter appeared. Some of its functions were partly replaced by the BioInstrumentation Advisory Council, which had been organized by the American Institute of Biological Sciences and produced pamphlets on various aspects of biotelemetry. The executive director of the council was Busser, the engineer who had led the American Electronics Laboratories' work on biotelemetry for ONR in the late 1950s and early 1960s.[68]

As the limitations of Cedar Creek's automatic system became clearer, the group began to focus its research efforts elsewhere. Mech, who had developed many of the field practices used by the group in the first few years of the automatic system, left Cedar Creek in 1966, when, as he recalled, Tester fired him because of a dispute over authorship. Mech subsequently used radiotelemetry to study the movements of wolves in northern Minnesota, where an automatic system along the lines of Cedar Creek's would have been of little use for tracking long-distance dispersal or for observing behavior in the field. By combining radio tracking with detailed studies of behavior and population dynamics, Mech soon became the leading expert on gray wolves. In 1967, Siniff told Eberhardt that the group was "spreading out a bit; hope it helps to find out a few new things." Cedar Creek biologists and engineers were now involved in a study of farmed pheasants at Carlos Avery, research on ducks in northern Minnesota, and a collaboration with Sargeant's Fish and Wildlife Service office in North

Dakota. Siniff himself was soon focusing his attention even farther afield. In 1967, George Llano, the director of NSF's Antarctic program, contacted Albert Erickson, a wildlife biologist at the University of Minnesota, about launching a study of Antarctica's seal populations. Siniff had worked with Erickson in Alaska's Department of Fish and Game before coming to Cedar Creek in 1964, and Erickson and Siniff were soon spending the austral summers tagging and counting seals out of McMurdo Station. They took advantage of the Cedar Creek Bioelectronics Laboratory's expertise to experiment with telemetry and other technologies in Antarctica, including the use of underwater cameras to record seal behavior. "Almost every trip south requires that an electronic type accompany the expedition," Siniff told Marshall in 1970.[69]

Although the AEC-funded irradiation study continued for four years, it was rarely mentioned in descriptions of Cedar Creek or the radiotelemetry system written for public consumption. In a 1968 report on Cedar Creek, for example, Marshall recognized that there was nothing simple about defining the meaning of the term "natural" on a site that had a long history of human uses, but he emphasized that the success of the radiotelemetry project, which "has attracted worldwide attention," depended on the Cedar Creek landscape's isolation from human influences. As Tester had promised, the total number of irradiated animals at large at Cedar Creek at any one time never exceeded six. By 1970, a total of ten raccoons, ten snowshoe hares, and six striped skunks had been irradiated and released. The researchers failed to recapture and kill only one irradiated animal out of the twenty-six. On February 13, 1970, three female snowshoe hares had been irradiated, tagged, and released with the intention of recapturing them a month later. About two weeks later, the radio signal from one of the hares, #264, disappeared. After a two-day search, Marshall was informed that the hare had been lost. Earlier that day two members of the museum group had shot a female and male pair in the area where the hare had been released, but neither turned out to be #264. By March 5, despite further searching, there was still no sign of her. Meanwhile other irradiated animals were being removed from Cedar Creek as planned. A yearling male raccoon irradiated the previous October, for example, was trapped and transported alive to a laboratory at the University of Minnesota to be examined, killed, and autopsied. Marshall was assured that no irradiated animals besides rabbit #264 had been lost. Ultimately, while the AEC-funded irradiation study had little effect on the Cedar Creek landscape, it also produced little of scientific value. While more irradiated raccoons than nonirradiated raccoons died in the month following treatment, those who survived showed no significant changes in home range, activity patterns, or breeding behaviors.[70]

As students who had carried out their thesis projects using the Cedar Creek system graduated and moved on to new institutions, they often continued to rely on the engineering expertise of the Cedar Creek Bioelectronics Laboratory. In effect, the laboratory began a small side business of providing transmitters and receivers to other research groups. The laboratory also built equipment for unaffiliated researchers, some of whom visited Cedar Creek to see the automatic radio-tracking system in action. These visitors came from as far afield as the Soviet Union, Lebanon, New Zealand, and Australia. In 1968, for example, the site's visitors included John Gibb, an animal ecologist from New Zealand with whom Marshall had worked during his Fulbright year, and Vladimir Sokolov, one of several Russian visitors whom Warner later remembered as "the most gentlemanly persons who ever visited our labs and automatic tracking system." Few of these visitors intended to set up automatic radio-tracking systems at their home institutions or field sites. On the contrary, they were sometimes concerned that the practices developed for the automatic system would not apply to their own work with less sophisticated equipment. In the fall of 1971, Tester had to reassure J. J. Lynch, an Australian visitor interested in using the software Siniff had developed, that "much of the data processed by these programs comes from portable telemetry equipment without any type of automatic data recording."[71]

Over the course of the 1970s, the Cedar Creek group itself became increasingly involved in projects away from the site. Graduate student thesis projects using the automatic system continued, and Tester experimented with using the system to quantify changes in daily activity patterns, but the most interesting research opportunities—and corresponding funding—were off site. In the mid-1970s Siniff became involved with a study of tigers in Nepal that involved two University of Minnesota graduate students, Mel Sunquist and J. L. David Smith. Like the Antarctic research, the Nepal project relied heavily on the expertise of the Bioelectronics Laboratory to build and repair equipment. In the late 1970s Siniff also began studying sea otters in Alaska's Prince William Sound under a grant from the Energy Research and Development Administration, which had taken over the research and development operations of the AEC as part of the establishment of the Department of Energy in 1975. As a result of environmental legislation—particularly the National Environmental Policy Act of 1969 and the Endangered Species Act of 1973—government agencies were required to determine the potential environmental effects of their activities. Reflecting this broader shift from Cold War to environmental funding priorities in the federal government, ERDA funded sea otter research because of concerns about the potential impact of oil spills on the small, vulnerable population of southern sea otters in California. An article about the Cedar Creek system published in 1982 noted that "in terms of actual biotelemetry research, recent years have witnessed a shift away from fundamental research to applied research oriented toward problem-solving. Consequently, the majority of research projects, although based at Cedar Creek, are conducted elsewhere."[72]

Technological developments also contributed to the Cedar Creek telemetry group's increasing focus on off-site research. Kuechle later attributed the group's turn to applied research in the 1970s largely to such factors: "We became more applied, I think, than we were earlier on, where we were answering specific problems, versus . . . when we first started. . . . You might consider it more basic research, I suppose, in developing the techniques, and once we had them developed, then, of course, we branched out into actually applied problems." Moreover, by the late 1970s, with the advancement of computer-based data collection and analysis systems, the film-based automatic telemetry system that had been at the cutting edge when Cochran designed it in 1963 had begun to seem cumbersome. Siniff recalled that "at that time computers had come so far . . . that we needed to update the system because the film reader was obsolete. . . . We needed to go directly from the towers into the computer."[73]

Funding for maintaining and updating the automatic system, however, had "started to dry up," as Tester later recalled. According to Siniff the group "started running out of money in the mid-70s" when the NIH training grant that had funded graduate students to work on the system was not renewed. In the late 1970s the university and the Minnesota state legislature provided small grants to update the computer systems at Cedar Creek, but these were stopgap measures. The proliferation of off-site, management-relevant projects and the increasing obsolescence of the automatic radiotelemetry system were gradually loosening the connection between the Bioelectronics Laboratory and the field site that surrounded it. The laboratory increasingly began to resemble the many small businesses that had been launched to provide biologists with radiotelemetry equipment since the mid-1960s. These businesses ranged from one-man shops such as Markusen's in Cloquet, which often specialized in a single type of animal such as upland game birds, to much larger operations such as Telonics, an Arizona-based company with roots in the aerospace industry, which offered telemetry equipment for everything from laboratory mice to gray whales. In the middle were companies such as AVM Instruments, which Cochran and his wife had founded in 1967. Although Cochran had "divorced out of" it after three or four years, it continued to operate under Barbara Cochran's leadership. Robert Hawkins, a biologist at Southern Illinois University whom Cochran had helped with deer-tracking studies in the mid-1960s, founded a telemetry company called Wildlife Materials with his wife Linda Hawkins in 1970. Telemetry companies sprouted outside of the United States as well, including Lotek in Canada, Biotrack in the United Kingdom, Televilt in Sweden, and Sirtrack in New Zealand. Radio-tracking gear and consultation was also available from other government and academic laboratories, such as the Denver Wildlife Research Center, which specialized in miniature radio tags that could be used to study crop pests, particularly small birds. In 1982, Kuechle noted that the balance had shifted over the previous decade from biologists working closely with engineers within a single research group, as at Cedar Creek, to biologists buying their equipment from commercial providers—a shift which Kuechle thought had slowed the pace of technological development.[74]

Even as suppliers of radiotelemetry gear proliferated, critics continued to raise doubts about the significance of the research that had been conducted with the technique. The substance of those doubts was not new, but it took on new urgency as radiotelemetry became more widely used. Robert Kenward, a British ornithologist who had begun using radiotelemetry to study raptors as a graduate student at Oxford in the early 1970s, recalled that the "honeymoon period was really finished about 1980, with the first really critical paper," by A. N. Lance and A. Watson, two Scottish biologists who participated in a conference on biotelemetry at Oxford in 1979. Echoing the anonymous NSF reviewer who had dismissed Marshall's research as "nothing more than natural history" in 1965, Lance and Watson accused researchers who used the technique of producing "mere descriptions of movements and activity, with no explicit hypothesis testing." Their opinion of radio tracking's unfulfilled promise was shared by a number of other participants in the conference, including Sargeant, who speculated that "more money and effort have been wasted on ill-conceived telemetry studies than on the use of any other field technique."[75]

By the early 1980s, the partnership of biologists and engineers that Kuechle believed was responsible for Cedar Creek's successes was under threat. As Kuechle later recalled, he and the other engineers, who were somewhat younger than Tester and Siniff, realized that when the biologists retired they would have a hard time justifying the continuation of the laboratory as an engineering-only operation. In 1981 Kuechle and his team—Ralph Schuster, Richard Huempfner, and Kathleen Zinnell—founded a private business called Advanced Telemetry Systems in the nearby town of Isanti while continuing to work at the Cedar Creek Bioelectronics Laboratory. The overlap produced some tensions. At a meeting in 1982, one member of the advisory committee—L. D. Frenzel, a Macalester College professor who had collaborated with Mech to launch his wolf-tracking studies in the late 1960s and regularly brought his students to Cedar Creek for field training—raised questions about "the use of Cedar Creek as a base for commercially building radio transmitters and receivers." David Parmelee, who had replaced Marshall as director of Cedar Creek in 1970, noted that he had informed the staff of Advanced Telemetry Systems in writing not to conduct any of their business on Cedar Creek property or to otherwise threaten the integrity of the research site in any way. Kuechle told the committee that he had resigned his position at the company and stayed on at the University of Minnesota for a few more years, but soon he too had left Cedar Creek to rejoin the company as its president.[76]

Tester later recalled that Kuechle's departure "was sort of the end of the automated system. That was shut down, and all the engineering staff was gone, and Cedar Creek became something else." The "something else" was a return to the kind of ecosystem research that had been dominant at Cedar Creek from the late 1930s to the early 1960s, when it had been overshadowed by rise of the "Museum men" and their telemetry work. In 1981, Cedar Creek was designated one of the National Science Foundation's eleven Long Term Ecological Research sites, with Tester and G. David Tilman, who had been appointed Cedar Creek's assistant director in 1980, as principal investigators. Under Tilman's leadership, the focus of Cedar Creek shifted away from wildlife telemetry and toward research on ecological succession. Although the LTER proposal included a budget for upgrading the automatic radiotelemetry system, Siniff later recalled that Tilman had little interest in pursuing it. The towers remained standing, as Marshall had worried they might when Warner first proposed installing them in 1962, but the system fell into disuse.[77]

In the 1980s, under the dual impetus of new funding sources for wildlife research and technological developments that had made radio tags more powerful, more reliable, and less expensive, the technology became seemingly ubiquitous. In 1985, one wildlife biologist pointed out in the *Wildlife Society Bulletin* that there were certain artifacts that identified wildlife biologists as a unique professional subculture, including the "tranquilizer dart-gun, telemetry animal collar, cannon-net, or animal tagging technology." As Tester later noted, wildlife telemetry was embraced by state and federal wildlife management agencies: "If you look at it from a practical standpoint, from a management standpoint—now you're a DNR [Department of Natural Resources] or a Fish and Wildlife Service, and your responsibility is managing habitat for a certain species—well, finally you know something about the requirements of that species. I mean, you always had sort of a broad feeling for it, but now there's some detail." The Fish and Wildlife Service, which had sat on the sidelines when the technique was being developed in the 1960s, prominently featured radio tracking in its promotional materials as a sign of the modernity and sophistication of its scientific work. In 1982, its Office of Public Affairs released *Role for Research*, a film that explained how science made it possible for wildlife to coexist with modern industry. Science produced "knowledge, for example, that permits birds of prey like the eagle to coexist with the chemicals which help make U.S. agriculture the most productive in the world." After describing FWS biologists' use of "tiny transmitters to track and study wildlife," the film displayed a close-up of one of the receivers designed by Cochran. "Nowadays Service scientists use the most sophisticated products of space-age technology," the voice-over explained, but "at the same time, they still rely on the traditional observation techniques of the field biologist to gather information."[78]

The visions of objectivity and transcendence over the messy contingencies of fieldwork that drove Marshall, Warner, and other wildlife biologists in the 1950s and 1960s to develop new techniques of remotely monitoring wild animals and their habitats helped convince the new Cold War–era institutions of federal science funding, such as NSF and the AEC, to support research that might earlier have seemed the exclusive domain of state fish and game agencies, the federal Fish and Wildlife Service, or private philanthropic organizations funded by wealthy sportsmen. In the popular media, they helped associate wildlife biologists with such icons of Cold War technology and modernity as missiles and satellites, providing them with a retort to accusations that wildlife biology had nothing to contribute to the practical problems of wildlife management. In the field, these visions drove the development of increasingly elaborate systems of observation, perhaps none more elaborate than the automatic radio-tracking system that Cochran designed for Cedar Creek, and they focused attention on the effects of observation itself on the animals and habitats under study. At both the Cloquet Forest Research Center and the Cedar Creek Natural History

Area, this attention led to disputes over the appropriateness of the interventions required or enabled by wildlife radio tracking, the results of which depended on the ability of researchers using radio tracking to maintain support from their federal sponsors.

After 1970, the rise of the environmental movement and new sources of support and demand for wildlife research would change this dynamic significantly. Increasingly, the Cedar Creek group would deploy its expertise to radio-track endangered, threatened, or otherwise problematic animals at multiple-use field sites rather than to simulate the effects of nuclear disaster on everyday animals at a nature reserve dedicated to science. Under these circumstances, the kinds of criticisms that Gullion had directed toward the Grousar project or that Lawrence and Marshall had directed toward the Cedar Creek group—criticisms based on the possibility that radio tracking would undermine the scientific value of a nature reserve—would fade in importance. And it would become possible for the Fish and Wildlife Service to celebrate its scientists' use of both "the most sophisticated products of space-age technology" and "the traditional observation techniques of the field biologist" without worrying about contradictions between the two.

CHAPTER TWO

The Poetry of Wilderness

Despite a brief flourishing of scientific research in the American national parks in the 1930s, the National Park Service's interest in science had never been strong. As late as the 1950s, for example, the iconic grizzly bears of Yellowstone or Mount McKinley had not yet been subjected to systematic study. In the late 1950s, however, the Park Service began encouraging wildlife biologists to apply some of the techniques of drugging, trapping, and tagging that they had recently developed to the study of grizzly bears and other wildlife in the parks. In 1957, the biologist Albert W. Erickson had reported at the North American Wildlife Conference his success in live-trapping and handling black bears in Michigan using the drugs pentobarbital sodium and succinylcholine chloride, a tranquilizer and a muscle relaxant. (This was the same Erickson who would later help Donald Siniff tranquilize and tag seals in the Antarctic.) Erickson's success inspired a number of other biologists to extend his techniques to other large carnivores, including grizzly bears. The Park Service also took notice. Heavily visited parks such as Yellowstone had been troubled by increasing human-bear conflicts as visitorship expanded dramatically in the postwar decades. Techniques for safely studying and handling bears held out the promise of resolving these conflicts without reducing either the number of bears or the number of visitors.[1]

The only region of the United States outside of Alaska with a significant number of grizzly bears was the northern Rockies, where they were protected in Yellowstone National Park and Glacier National Park. In the late 1950s, Park Service naturalists began discussions with John Craighead, head of the Montana Cooperative Wildlife Research Unit, about studying the region's bears. Although Craighead had no particular experience studying large carnivores, he was a rising star in wildlife biology; he and his twin brother and close collaborator, Frank Craighead, had developed a number of innovative techniques for studying raptors, waterfowl, and other wildlife. In November 1957, Gordon Fredine, the Park Service's chief biologist in Washington, suggested that Craighead conduct a study of the entire grizzly population of the northern Rockies, but Craighead's preference was for a narrower study focused on the parks. That preference became even stronger after Craighead visited Yellowstone's Trout Creek garbage dump with one of the Park Service naturalists in the fall of 1958 and witnessed the high concentration of bears feeding there. Craighead immediately realized the research opportunity provided by the dump, even though, as he told Fredine after the visit, any data gathered would have to be interpreted in light of the "artificial conditions." The Craigheads and their collaborators would later write that they had selected Yellowstone for the study because it had a "wild grizzly population sufficiently free of artificiality that fundamental biological data could be obtained an on a quantitative basis." But it was the Trout Creek dump and the "artificial conditions" for science that it provided in combination with the park's natural conditions—not the latter alone—that convinced John Craighead and Frank Craighead to study Yellowstone's grizzlies.[2]

Despite strong interest in the study among Park Service leadership in Washington and at Yellowstone, the Park Service was unable to provide the Craigheads with any funds. Indeed, it was the lack of funding for research within the Park Service that had led Fredine and others to reach out to the Craigheads, who could draw on other sources of support. The Craigheads initially turned to the Wildlife Management Institute, which provided a small exploratory grant for the first field season, and to the National Geographic Society's Committee for Research and Exploration, which had supported much of their earlier work. The Craigheads began experimenting with techniques for handling bears in 1959 with the assistance of Maurice Hornocker, one of John Craighead's graduate students, and Wesley Woodgerd, a biologist with the Montana Fish and

Game Department. By July 1959 they had successfully trapped seven bears and immobilized them using the same muscle relaxant that Erickson had used. The Craigheads focused their work on bears gathering at the Trout Creek dump, who could be easily lured into large traps made out of culvert pipes with bait such as honey, bacon, or pineapple juice. They also used dart guns to fire drug-loaded syringes at free-ranging bears. In addition to being weighed, measured, and photographed, the bears were given small ear tags, large plastic ear streamers, and identifying tattoos.[3]

Because the Craigheads were using a muscle relaxant for immobilization, the bears were awake during these procedures. In an article written for *National Geographic* after the first season of research had been completed, the Craigheads reported a conversation with one park ranger who was worried that the bears might retaliate against the researchers who had treated them so roughly. The Craigheads joked that "each time we handle a grizzly, we slip into ranger jackets and put on ranger hats," but they also wrote, more seriously, that they had no evidence the bears "remember or associate us with an unpleasant experience." John Seidensticker, however, who joined the Craigheads as an undergraduate field assistant in the mid-1960s, later recalled that it was difficult to catch a bear the second time: "The bears really, really figured you out. . . . They didn't pay any attention to [the Craigheads'] green station wagon to start with, but after a while they seemed to skirt way around it, and they'd figured out that bad things happened out at the car." Because the researchers were using only a muscle relaxant, "the bear is wide awake, just fully immobilized, and so he's fully awake to what's going on."[4]

By July 1959, park naturalists and rangers had begun to notice tagged bears around the park. In August Yellowstone superintendent Lemuel Garrison sent a memo to all park staff describing the Craigheads' study, including its use of highly visible colored markers, and urging them to cooperate. He also issued a press release noting that it would "be necessary to capture and mark individual bears so they can be definitely recognized in the field and their movements and activities recorded." By the fall, the Craigheads later reported, "Reports on marked grizzlies were trickling in from fishermen, campers, hikers, and park rangers." During the first field season the Craigheads captured and marked twenty-seven bears, most of them in the area around the Trout Creek dump. Three bears died from accidental drug overdoses, but the Craigheads argued that the use of drugs had resulted in a net gain in the population, since a number of troublesome bears who might otherwise have been killed were instead safely captured, tagged, and relocated. John Craighead told Garrison that he hoped the study would reveal a way that "bears could be kept out of the garbage dumps and at the same time discouraged from frequenting the campgrounds."[5]

The Craigheads' *National Geographic* article appeared in the spring of 1960 under the title "Knocking Out Grizzly Bears for Their Own Good." It emphasized the danger and excitement of the research, the Craigheads' benevolent power over their bears, and the importance of ear tags and other markers for transforming each bear into a "recognizable individual" who could be repeatedly observed. Though the Craigheads had not mentioned radio tracking in their original research proposal to the Park Service, by the time they wrote the *National Geographic* article they had come to believe that the technique would be central to their study of Yellowstone's grizzlies. To ensure the survival of the population, they wrote, "We needed to learn more about the movements of grizzly bears, the size and character of their ranges; about population structure and numbers, breeding age and frequency of breeding, litter size at birth, and infant mortality. We wanted to study their condition during the deep sleep of winter, to determine their average life span, and many other things." Radio tracking was essential to accomplishing these goals, they told their *National Geographic* readers. The technique, which they planned to begin testing in the 1960 field season, promised to allow them to follow individual grizzlies from the dumps to the rest of the park and beyond and to locate bears at night and in densely forested terrain—something that visual tags could not do.[6]

While John Craighead coordinated the study as a whole and served as the primary liaison to the Park Service, Frank Craighead took the lead in developing radio-tracking equipment. For technical assistance, he turned to Hoke Franciscus, a childhood friend and ham radio enthusiast from Pennsylvania, and to Joel Varney and Dick Davies, engineers at the Philco Corporation's avionics division in Palo Alto. The development of the equipment was largely funded through a grant proposal that the Craigheads submitted to Sprugel's Environmental Biology Program at the National Science Foundation in early 1960. In March, Frank Craighead attended the telemetry session at the meeting of the Wildlife Society that was organized by South Dakota biologist Jack Seubert. He subsequently became one of the founding members of the "radio tracing" subcommittee of the Wildlife Society's Techniques Committee, along with William Marshall and Lowell Adams.[7]

Once the Craigheads had demonstrated that the park's bears could be safely trapped, tranquilized, tagged, and translocated, the Yellowstone administration began to incorporate the new techniques into its grizzly bear and black bear management practices. In the spring of 1960, Yellowstone's chief naturalist, Robert M. McIntyre, told John Craighead he was enthusiastic about the technique's potential to improve bear management, and the Park Service's regional chief of interpretation told Garrison that he thought the new techniques would be useful for managing "troublesome black and grizzly bears." The Yellowstone administration was also supportive of the Craigheads' efforts to publicize their research. In May 1960, McIntyre congratulated John Craighead for having the *National Geographic* article published, noting that it might help them raise additional funds for their research, which would ultimately benefit the park.[8]

During their second field season in Yellowstone, the Craigheads trapped forty-seven individual grizzlies, of which three died from drug overdoses. The rest of the bears were successfully captured, measured, marked, and released. By the end of 1960, a total of seventy-seven grizzlies had been marked since the beginning of the project, which by the Craigheads' rough estimate amounted to perhaps one of every four grizzlies in Yellowstone. In the summer of 1960, the Craigheads also began testing dummy radio collars for durability. The radio-tracking equipment that they had contracted Philco to build and the necessary radio permits from the Federal Communications Commission became available in October, and late that month they captured a grizzly sow in the hope of attaching their first functioning radio collar. Finding, however, that the bear's prewinter weight gain had made her neck too thick for the collar—a full twenty-nine inches in circumference—they instead used her as a mannequin, testing the impact of various placements and orientations of the collar on the strength of the signal emitted from the collar. While the tests left the Craigheads optimistic about the transmitting equipment, the details of a "trailing system" that would allow them to locate the far-ranging grizzlies in rough and often inaccessible terrain were still unclear.[9]

While the Craigheads worked the kinks out of their radio-tracking system, plans to tag grizzlies were also moving forward in Alaska, the only other region of the United States with a significant population of the bears. In December 1960, Adolph Murie received a letter from the chairman of the Arctic Institute of North America asking for his evaluation of a proposal from biologist Frederick Dean to radio-tag grizzlies in Mount McKinley National Park. Murie was a Park Service biologist whose studies of coyotes and wolves in Yellowstone and McKinley in the 1930s and 1940s had helped change the Park Service's approach to predator control. Inspired both by Murie's research and by new techniques that Erickson, the Craigheads, and the Alaskan biologist Will Troyer had pioneered for handling bears, Dean had begun conducting quantitative observations of McKinley's grizzlies in 1957 with the intention of eventually trapping and tagging them. Although Murie had supported Dean's earlier work, he was disturbed by the idea of tagging what he called "wilderness wildlife." In his evaluation for the Arctic Institute, written from his summer home in Jackson Hole, Wyoming, just south of Yellowstone, Murie described the tagging of grizzlies as a "newly-tried but acceptable scientific technique" from "the standpoint of technical procedure" but one that was for other reasons completely inappropriate for use on McKinley's bears.[10]

Those reasons fell into three broad categories. First, Murie argued, there was no need for the knowledge that would be acquired. In his proposal, Dean had described the current state of biological knowledge of McKinley's grizzlies as "meagre," an epithet that clearly upset Murie, who had been slowly accumulating observations of the population for more than a decade. The threats to grizzlies were already clear, Murie suggested: loss of wilderness, livestock grazing, and uncontrolled hunting. Rather than more research, what was needed was "more esthetic writing about grizzlies in wilderness environment in order to create more sentiment for saving them." Second, wildlife tagging would undermine the wilderness character of the park. It was particularly threatening to "the McKinley wilderness, which we are trying to maintain at a higher and purer spiritual and esthetic level than is possible in such parks as Yellowstone and Yosemite." The grizzlies in McKinley were still "*wild* grizzlies," unlike the bears "contaminated" by garbage that the Craigheads had been tagging in Yellowstone. Finally, the proposed study risked alienating many of the Park Service's strongest supporters. Even if it were officially approved, Murie warned, "it is entirely possible that public protest might force the termination of the marking phase of the study." Perhaps to indicate from which quarter such protest was likely to come, Murie included a statement from his brother Olaus Murie, a wildlife biologist who had recently served as the president of the Wilderness Society. Olaus compared Dean's proposal to an earlier proposal to establish a scientific station near the top of Mount McKinley, which had been blocked by wilderness activists and the park administration "because of its impact on the esthetics of a national park, even though the proposal had scientific merit." Tagging, he argued, would have a similarly drastic effect on park aesthetics without meeting any urgent conservation need.[11]

While Robert Mason, the secretary of the Arctic Institute's Research Com-

The ear tags and radio collars that John and Frank Craighead attached to Yellowstone's grizzlies in the 1960s were brightly colored and highly visible, both to the researchers and to park visitors and staff. (Courtesy of the Craighead Archives: John and Frank Craighead/Craighead Environmental Research Institute)

mittee, told Adolph Murie that his "argument for protecting the grizzlies in McKinley National Park, even at the expense of limiting investigation of them, is compelling," the Arctic Institute's decision ultimately had nothing to do with wilderness ideals. As Mason later explained to Murie, "The Institute, as far as I know, has no expressed policy regarding protection of wilderness values versus conduct of field research." Instead, the committee had based its decision not to fund Dean's study on "the size of Dr. Dean's request," "the question of whether this kind of research should be sponsored by state or local agencies," and the fact that the research was "a little too far outside the area of our most serious interests to be considered." Park Service leadership was even less sympathetic to the Muries' concerns. Despite "some reluctance to approve the study" among his staff, regional director Lawrence Merriam told Olaus Murie in early January 1961, before the Arctic Institute had come to a decision about Dean's proposal, "It was felt that the values to be derived from the research project were of such importance that the benefits outweighed the undesirable aspects." Moreover, it was expected that "judicious capture methods and the use of inconspicuous marking devices" would make it possible to carry out the study without any detrimental effect on the bears and without the visiting public even being aware of its existence.[12]

Rather than treating radio collars as something that could or should be hidden from the public, Yellowstone superintendent Garrison treated them as welcome signs of the modernization of Park Service bear management. By July 1961, Garrison was promoting the Craigheads' bear-handling techniques to administrators of other natural areas who had heard about them through scientific conferences, the popular press, or word of mouth. Bears who frequented campsites or begged for food at roadsides could be trapped, immobilized, tagged, and transported to remote areas of the park and then monitored for recidivism. The technique seemed safe for both bears and men; Garrison told one inquirer that park staff had not yet "killed a black bear that we didn't want to." In the fall of 1961 the Yellowstone administration issued several press releases about the park's use of immobilizing drugs and "modern devices" of wildlife research such as radio collars. It also welcomed a National Geographic television crew to the park to film the Craigheads' first real use of radio tracking to study the park's bears. In August 1961 the Craigheads tested the collar on a human subject, and in September they collared two grizzlies. The first bear they collared was Number 40, a female they called Marian after the wife of one of the Philco engineers.[13]

For the Craigheads, radio tracking was both a way of obtaining a panoramic view of the grizzly population of Yellowstone and a way of focusing in on a particular bear or a specific aspect of grizzly life history. As they later wrote, radio tracking provided two complementary ways of "extend[ing] the range of man's observational powers." One was by allowing data to be "gathered from a distance with a minimum of time, effort, and man-power." But "an even greater asset of a tracking system is its capability of putting a scientific observer in a predetermined position where he can observe a particular instrumented animal and reason about and interpret what he observes on the spot." The Craigheads monitored radio-tagged bears from their laboratory at Canyon, from a field station on a hill eight miles to the south, from automobiles and small planes, and on foot. Remote signals indicating a bear's location or activities helped them determine when to move in for close observation. Frank Craighead later noted that one of the advantages of radio tracking was that "for the first time in our study of the grizzly, we had the advantage of surprise on our side. . . . [O]ur purpose was not to avoid the grizzlies; the point was to get close and observe them without being detected." This close observation gave the Craigheads a rich sense of the bears as individuals. Like Marian, many of the bears they tagged received names as well as numbers: Pegleg, Scarface, Cutlip, the Fifty-Pound Cub, the Rip-Nose Sow, Loverboy, Notch-Ear, Old Short Ears, Shorty, the Grizzled Sow, and the Sucostrin Kid—the latter a yearling male with an unusually high tolerance for the immobilizing drug. Frank Craighead later described the relationship between scientists and bears as one of "mutual trust, and perhaps respect," or at least tolerance.[14]

The Craigheads' work was characterized by a combination of wilderness advocacy and technological enthusiasm that is best captured in Frank Craighead's description of the tagging and tracking of Marian, bear Number 40, in his 1979 book *Track of the Grizzly*:

> Beep, beep, beep, full of portent and meaning, the repetitive metallic pulse came in loud and clear on the crisp fall air. The sound had nothing of wildness about it. No deep primitive instinct of the chase stirred in us at the sound, nor did it evoke a feeling of oneness with nature. Yet this beeping coming to us in the vastness of Hayden Valley thrilled us as few sounds ever had. The vibrant pulsing signal, though new to the Yellowstone wilderness, told us that we were in communication with the grizzly we identified as bear Number 40, just as surely as the distant honking told us that the Canada geese were on the wing. But the beep was

more specific than the honk of the goose or the guttural caw of the raven, for it emanated from one particular grizzly bear somewhere within the three thousand square miles of the park. Hearing this sound meant that we were monitoring the first free-roaming grizzly sow to be tracked by radio. . . . Number 40's debut as a free-roaming electronic instrument of science took place that day.

For the Craigheads, radiotelemetry complemented older methods of tracking rather than replacing them. On one occasion during the first year of tracking, the team identified footprints that they thought had been left by Marian, and they followed those tracks until they were able to pick up her radio signal.[15]

In contrast, the Muries rejected the idea that space-age technology, especially the kind that involved hands-on manipulation and permanent marking of "wilderness wildlife," could be used to carry out research without losing a vital connection to the animals and their habitat. Their resistance to new techniques was reflected in Olaus Murie's attitudes toward research in the Arctic National Wildlife Refuge on Alaska's North Slope. At a conference on Alaskan science in 1961, Olaus Murie called for the refuge to be devoted to "basic scientific research, with the least possible equipment. It should be for the kind of scientific study based on thinking, based purely on close observation, trying to understand the relations among various animal forms and the changing environment. We need to understand more, to interpret honestly, what we see in wild country." Close observation and honest interpretation, not gadgetry, were the keys to scientific advancement. Privately, Murie told the executive director of the Wilderness Society that he was pleased that the Fish and Wildlife Service budget for research in ANWR had been cut, limiting the Service's activity there. Given FWS's managerial orientation, Murie believed any research by the agency was likely to be highly invasive.[16]

In the eyes of the Muries, the Park Service's management philosophy was preferable to that of the FWS, but its implementation often fell short of the ideal. In December 1961, Olaus Murie wrote to Merriam again to criticize the Park Service's handling of a variety of issues, including the grizzly bear research being conducted by the Craigheads in Yellowstone and by Troyer on Kodiak Island. In his reply, Merriam disputed Murie's characterization of one of the researchers as a "proud hunter with a gun"—an image Murie believed undermined fundamental park values—and denied being aware "of any resentment being generated by the current studies of grizzly bear behavior." He defended the tagging of bears in terms of the knowledge gained and belittled the aesthetic

Although the Craigheads carried out much of their tracking of grizzlies from fixed antenna stations and a central laboratory, they also used portable receivers to locate grizzlies in the field for direct visual observation. (Courtesy of the Craighead Archives: John and Frank Craighead/Craighead Environmental Research Institute)

concerns of a nature photographer that Murie had mentioned: "It is our understanding that the studies will provide essential data looking toward the perpetuation and well-being of the species. Perhaps the ear tag in your photographer friend's picture is justified by the fundamental knowledge gained."[17]

The Muries were not entirely alone, however, in their concern about tagging. Even Dean was more sympathetic to such concerns than he had first seemed. As

Murie later told the director of the Park Service, Conrad Worth, Dean "agreed, indeed suggested himself, that certain kinds of scientific work would not be suitable in a natural area, because of the disturbance it would cause" when Murie met with him in McKinley in the summer of 1961. Exactly which kinds of scientific work were unsuitable, however, remained open to debate. In February 1962, Richard Prasil, a member of Merriam's regional office, requested Adolph Murie's views on "the need for an intensive grizzly bear study through the use of anesthesia as a means for live capture for tagging and/or subdermal placement of radio transmitters." Dean later recalled that he had proposed subdermally implanting the tags precisely in order to address the concerns about the aesthetic impact of external tags and collars that Murie had raised. This change in procedure, however, did not assuage Murie's concerns. Although tagging would undoubtedly provide some new information, he told Prasil, his own work had resulted in "a rather full understanding of grizzly ecology in the park." In terms of the conservation needs of the grizzly, there was "no urgency" for further work. Moreover, because bears did not concentrate at garbage dumps in McKinley, tagging would be logistically much more difficult there than in Yellowstone. Finally, Murie's discussions with people inside and outside the Park Service had revealed "very strong feelings against not only the proposed McKinley tagging project but also the one in Yellowstone." In a second memo sent a few days later, Murie admitted that "it is true that in a highly publicized Yellowstone study the grizzlies carry tassels, etc. It is also true that when we think of Yellowstone grizzlies we do not think of wilderness animals, but rather of radios, anesthetized bears, and general manhandling." Yellowstone's bears were contaminated; McKinley's should remain pure.[18]

Adolph Murie believed that one of the reasons such unacceptable proposals were under consideration was that the Park Service lacked a strong scientific staff of its own. On the same day that he wrote one of his lengthy memos to Prasil opposing the radio tagging of grizzlies, Murie also responded to an internal Park Service report titled "Comprehensive Natural History Research Program for National Parks," which proposed several means of strengthening the Park Service's science program. To Murie, the report seemed to recommend relying too heavily on outside researchers, when instead what the Service needed was its own "corps of outstanding field biologists whose leadership in wilderness wildlife conservation would be nationally recognized." As past experience had shown, not all outside researchers were "in sympathy with park policies," nor were they able to provide the Service with institutional continuity. Murie also criticized the report's focus on "efficiency" and "research teams." The language of the "efficiency expert" seemed "a little incongruous in the woods," he wrote, especially since the benefits of field research were often intangible: "Contribution of a phrase as poignant as 'Reverence for Life' might, for example, justify several years of hire." The fundamental aim of his own scientific work, Murie claimed, was just such an intangible result: to "promote a sympathy for wildlife and faith in the basic philosophy of the National Park Service."[19]

Uplifting to the Human Race

The Muries' concerns about the role of outside scientific advisers in shaping Park Service policy were confirmed in their eyes by an influential report on wildlife management in the national parks produced by an ad hoc advisory committee in the spring of 1963. The committee was headed by A. Starker Leopold, an influential wildlife biologist at the University of California, Berkeley. According to park historian Richard Sellars, the Park Service commissioned the report in response to a "crisis in public relations" arising in large part from controversy over the management of Yellowstone's elk herd, particularly the question of whether sport hunting should be used to reduce the burgeoning population. As Sellars writes, the Leopold Report, as it came to be known, "inspired a patriotic, ethnocentric goal—to maintain the landscape remnants of a pioneer past as they were 'when first visited by the white man' or when 'viewed by the first European visitors.'" This "wilderness *pastorale*" appealed to many Park Service employees, but it also implied something with which many were less comfortable: namely, that these "vignette[s] of primitive America" be maintained or re-created through active management by biologically trained Park Service personnel. In line with this recommendation, the report recommended against allowing public hunting in the parks, while also recommending a variety of ways in which the Park Service could and should manipulate park landscapes.[20]

Inasmuch as the Leopold Report opposed public hunting, the Muries supported its position. They believed hunting was valuable in principle, but they had a dim view of most of its practitioners in modern America. For Olaus, as for Adolph, the risk of allowing hunters into the park was not that it would harm individual elk or the elk population but that a horde of undisciplined hunters would undermine the wilderness experience and thereby an American culture founded on the European encounter with the raw nature of the New World. That culture was already under threat. As Olaus Murie explained to

David Brower, the executive director of the Sierra Club, in June 1960: "In all phases of American life we stress the importance of size, numbers, grandeur, masses—rather than quality of human experience. . . . I feel that the harm done in modern American hunting is not so much to the animals involved as it is to the human spirit, and so to our declining culture." The Muries' opposition to public hunting in the parks, like their concern about handling of "wilderness wildlife," harked back to Victorian-era humane concerns about the effect of cruelty on the souls of those who committed or witnessed cruel acts. As Adolph Murie explained to the head of the National Parks Conservation Association in January 1963, "There are scientific and other values in parks, but in my opinion the most fundamental values are in the realm of the esthetic and the spiritual. Our park ideals are an expression of the best in us. Our better instincts are given free play, and we have an opportunity to show tolerance and kindness toward our fellow creatures. This, I believe is uplifting to the human race."[21]

Though they approved of the Leopold Report's opposition to public hunting, the Muries were dismayed by its support for aggressive manipulation of park landscapes and animal populations, which in many cases would depend on the kinds of techniques pioneered by the Craigheads in Yellowstone. A few months before his death in the summer of 1963, Olaus Murie told one colleague that he was disappointed with much of what passed for "modern science. So many of the modern 'wildlife management' scientists are spending so much of their time flying about in airplanes, counting animals, or marking them; they don't get down and live with the animals and learn how they live." A second report on science in the parks that was then underway by an ad hoc committee assembled by the National Academy of Sciences was more to the Muries' liking. While the Leopold Report had focused on preserving landscapes as they were believed to have looked at the moment of European discovery, the National Academy committee report, which was completed five months later, argued that the preservation of ecological and evolutionary processes should be the Park Service's primary goal. It defined parks as "dynamic biological complexes" and suggested keeping management to a minimum, although, like the Leopold Report, it also recommended increasing the Park Service budget for science.[22]

Although Sellars describes the two advisory reports as "a kind of ecological countermanifesto that marked the beginning of renewed efforts to redefine the basic purpose of national parks," contemporaries saw fundamental philosophical differences between the two. Adolph Murie strongly favored the National Academy report over the Leopold Report, as did the Wilderness Society, which was then in the last stages of a long campaign for a federal wilderness act. An editorial in the society's magazine, *Living Wilderness*, described the Leopold Report as "a serious threat to the wilderness within the National Park System and indeed to the wilderness concept itself." The editorial noted that the report's stated goal of reproducing the landscape as the white man first saw it was as idealistic as the Wilderness Society's goal of letting wilderness areas develop according to natural processes without intensive human management: "Wilderness preservation is, indeed, itself a fancy, a human concept, but its inspiration is to use 'skill, judgment, and ecological sensitivity' for the protection of some areas within which natural forces may operate without man's management and manipulation." Hands-off wilderness protection to preserve natural processes and hands-on wilderness management to preserve natural states were both unattainable ideals; the question was which ideal should guide a necessarily imperfect practice.[23]

While Leopold continued to push the promanagement approach outlined in his committee's report, Murie tried to convince Park Service leadership and other wilderness activists that his approach was "extremely dangerous to park policy." Early in 1964, in a commentary titled "A Plea for Idealism in National Parks," he pointed out that Leopold's seemingly idealistic vision of the parks would entail intensive management by man in order to freeze wilderness in time. If this vision were adopted, "the natural forces, the natural ecological processes shall be halted." In a speech to park superintendents, Leopold had recently suggested that roadside clearing of trees was justified in the service of producing "vignettes," an argument that struck Murie as "the most extreme anti-park policy statement I have yet encountered." Leopold's approach, Murie warned, "would be appreciated by the 'wildlife managers' in that it would give them a big field for plying their various techniques." The fact that it was being taken seriously was all the more reason for the Park Service to develop its own internal science capacity. If the Park Service had its own experts who believed in the national park ideals, it would not have to rely on outside experts who so evidently did not. After learning of Murie's views, Howard Zahniser, director of the Wilderness Society, told him he shared his concerns: "Whether we are contending with a real romantic nostalgia or with a rationalizing of an urge to introduce massive management we should be doing all we can to point out, as you so well do, the meaning of all this."[24]

The Craigheads considered themselves just as much wilderness advocates as the Muries, but they believed that wilderness could not be left to its own devices;

it had to be carefully managed with the help of the best science available. Their study of Yellowstone's grizzlies was an example of the kind of scientific work that would support appropriate wilderness management. In 1964, in their annual report to the Yellowstone administration, the Craigheads recommended that Hayden Valley, site of the Trout Creek dump, as well as the entire area within Yellowstone's "Grand Loop" of central roads be maintained in as close to a wilderness state as possible. The Yellowstone administration forwarded the report and a paper on biotelemetry that the Craigheads had presented at the Instrument Society of America in September 1963 to the Midwest regional office, which sent copies to Park Service director George Hartzog's office in Washington. The regional office asked the Yellowstone administration to "carefully evaluate the papers" and report back to the office "on how the recommendations will effect [*sic*] long-range development and management plans for the Park." After reading the report, Yellowstone park biologist Walter Kittams told John Craighead that his "concise Management Suggestions are down-to-earth, a notable contribution from your studies and worthy of our best efforts."[25]

While the Wilderness Society was sympathetic to the Muries' concerns about management, it also gave voice to those such as the Craigheads who thought that unmanaged wilderness was a misguided fantasy. In the spring of 1964, the *Living Wilderness* reprinted an article by John Craighead that argued that "by applying present knowledge and seeking more, through research, we can use our wilderness heritage to the full and still retain it essentially as we know it today. We cannot just visit and otherwise leave it alone. Eventually we must understand it and manage it ecologically just as we do any area of land where man's impact is felt." When the Wilderness Act became law in 1964, it designated scientific research as one of the "public purposes" of wilderness areas, along with recreational, scenic, educational, conservation, and historical uses. In the years after the passage of the Wilderness Act, the focus of wilderness advocacy would shift from how to gain federal protection of wilderness areas to how—and whether—to manage areas designated under the act. The Craigheads were among the most prominent advocates for a hands-on approach. In the fall of 1965, the *Living Wilderness* noted that Frank Craighead had become president of the new Environmental Research Institute, which would focus its attention on problems of "wilderness management."[26]

While debates over "management" versus "preservation" continued within the Park Service and among wilderness activists, the Park Service sought to strengthen its internal scientific capacity, something both the Leopold Report and the National Academy report had recommended. In 1964, Hartzog created a new Division of Natural Science Studies and appointed Sprugel, formerly the head of NSF's Environmental Biology Program, as the Park Service's chief scientist. As the Park Service's ambitions for its own scientific programs increased, so did tensions with the Craigheads over their research in Yellowstone. Superintendent Garrison remained a strong supporter of the Craigheads through the end of his term in 1964, when he was replaced by John S. McLaughlin, but negotiations to renew the agreement to work in the park were becoming increasingly difficult.[27]

One particularly tense subject of negotiation was a proposal to radio track elk and other "big game" animals. Like Tester in Minnesota, the Craigheads had responded to Schultz's call for proposals at the telemetry session at the North American Wildlife Conference in March 1963 by submitting an application to the AEC for funding. While the Craigheads had not proposed irradiating Yellowstone's elk—a definite nonstarter, no matter one's stance on wilderness management—the new project did raise questions for the Yellowstone administration about how long the Craigheads planned to continue their operations in the park and how extensive those operations would become. When a renewed agreement was finally reached in November 1964, John Craighead described the negotiation as a "long siege." Subsequent negotiations were even more difficult. The Craigheads accused the Park Service of using restrictive contract clauses to gradually take control of the grizzly project, while the Park Service began to suspect that the Craigheads, rather than conducting the time-limited and management-relevant study they had initially been invited to carry out, had decided to make Yellowstone their own permanent laboratory and training ground for graduate students. In June 1965, concerned about "unseemly terms and language" in a new contract that required the Craigheads to provide information to the Park Service, John Craighead assured McLaughlin "that my only interest and those of my colleagues is to conduct research on vital problems in Yellowstone National Park and to make this information available to the National Park Service." The fact that the Park Service was now providing a small portion of funding for the project did not give it the right to take control of the Craigheads' work; in fact, Craighead wrote, they had accepted the Park Service's funding mainly as a sign of good will and interest in cooperation. Now the Park Service seemed to be trying to use that minuscule funding as a justification to "completely engulf our work into a rigid service program and to take credit for our accomplishments without even the courtesy of treating us as fellow colleagues."[28]

While the Craigheads struggled with the Park Service for control of grizzly research in Yellowstone, Dean was still seeking permission and funding for a grizzly-tagging project in McKinley. Even after Adolph Murie retired from the Park Service at the end of 1964, he continued to oppose the proposal; in fact, now that he was no longer a Park Service employee, he felt free to express his opposition more forcefully. On January 1, 1965, he wrote to Freeman Tilden, author of *The National Parks: What They Mean to You and Me*, to express his concern about two developments in McKinley that threatened the park's wilderness character. The first was a plan to expand three and a half miles of the McKinley park road in continuation of a project that had been under way since the 1950s. Unfortunately, Murie told Tilden, none of the three McKinley superintendents who had supervised the expansion of the road were "wilderness oriented." The threat of road building paled in comparison, however, to a second development: Dean's grizzly-tagging proposal. "The flagrant super-road construction and other developments are unfortunate—they make an intrusion far more destructive than need be," Murie explained to Tilden. "But it seems to me that marking grizzlies, milking them, branding them like cattle, laboratoryizing and de-mystifying them takes us almost into another realm. For we are destroying the very essence of the poetry of wilderness." Not only was such research unnecessary for the conservation of the grizzly, Murie argued, but there were ample areas outside of "sacred" parks and wilderness areas where scientists could "play with the marking technique" if they so desired. In its decision about whether to allow tagging in McKinley, the Park Service was, in Murie's opinion, "on trial."[29]

Murie also conveyed his concerns to Sprugel, whose background as one of the funders of the Craigheads' radio-tracking work made him an unlikely sympathizer. Murie told Sprugel that the current superintendent of McKinley was opposed to tagging the park's bears, but "projects of this kind are not easy to stop and perhaps it has been passed up the line." Tagging grizzlies and other hands-on scientific manipulations would, he wrote, "put a blight on this entire northern sanctuary" without providing any information vital to conservation. More generally, he warned, Leopold and other professional wildlife managers had been allowed far too much influence over park policy. Sprugel's reply was noncommittal. Although his staff were aware of Dean's long-standing interest in studying McKinley's bears, no one had seen an official proposal for the study, and, to the best of their knowledge, no one in the Park Service had officially approved it. Sprugel pointed out that it was impossible to guarantee that Dean's study would not be approved in the absence of a proposal detailing "the objectives of the research or the techniques he proposes to employ to achieve these objectives," but he added that, "even if granted, approval for a venture of this nature will not be forthcoming until the merits and undesirable features of the undertaking have been carefully evaluated and weighed."[30]

One of Murie's potentially strongest arguments against radio tagging was that park administrators already had enough information about the lives of grizzlies to protect them. This argument was weakened, however, by the fact that Murie had not yet published the results of his observational research on McKinley's grizzlies. Prasil, the regional office staff member who had requested Murie's opinion on tagging in 1962, urged Murie to publish the results of his grizzly research as soon as possible if he hoped to forestall the use of the technique in McKinley. Sprugel had recently dismissed the protests of an Anchorage-based nature photographer by arguing that the differences of conditions between Yellowstone and McKinley justified tagging bears in both locations, and although Prasil agreed with Murie that "tagged bears in McKinley would be a jarring sight and a spoliation of the wilderness scene," he did not believe that such arguments would suffice in rebutting Sprugel's claims. In previous correspondence, Murie had admitted that tagging might be justified if it would produce information necessary for conservation, but he now told Prasil that he would oppose tagging regardless of the circumstances. This was a far more radical stance than Murie had taken just a few years earlier, one that he recognized as such. Opposition to scientific practices posed a special challenge, Murie told Prasil, since "many people, especially scientists, feel that research is a sacred word, so that we cannot oppose what is done in its name. Therefore, it takes special dedication, special effort, special caring, to oppose a research intrusion."[31]

Although Murie was unable to win over Sprugel, he found more sympathetic ears within the McKinley administration. After 1965, Dean gave up on attempting to radio-tag McKinley's grizzlies and turned his attention elsewhere. In Yellowstone, science had trumped aesthetics. The historian Alice Wondrak Biel notes that Yellowstone superintendent McLaughlin responded to one visitor's complaint about seeing a bear with tagged ears during a visit in 1965 by explaining that the park had decided to "sacrifice natural appearance" for the sake of research. But McKinley's stronger claim to "wilderness character" and its less pressing management problems made such a sacrifice seem unnecessary.

The management plan for Mount McKinley National Park that was completed in the fall of 1965 celebrated the park's wilderness characteristics and asserted that the park would "continue to serve its highest purpose if the major portion is included" in the federal wilderness system established by the Wilderness Act. Under the heading "Resource Conservation Objectives," the plan recommended limiting "the collection of natural history specimens and historical artifacts to those necessary for display or for study for authentic presentation of the interpretive theme." This relatively hands-off approach would persist into the 1970s. Although Murie continued his observations of the park's bears each summer, in 1968, when Prasil reported the results of an aerial survey of wildlife in McKinley, including some incidental observations of grizzlies, the surveys on foot and by air that Dean had conducted in 1958 were the most recent grizzly population estimates available.[32]

By 1966, Park Service leadership was aware that scientific research on park wildlife might attract criticism from both inside and outside the Park Service, and not just from wilderness purists such as Murie. In the summer of 1966, the acting superintendent of Yellowstone, J. M. Carpenter, told one outside researcher that his proposal for physiological studies on the park's bears was unlikely to be approved, since "the accessible portion of the year for wildlife studies largely coincides with the visitor season, and we cannot emphasize research studies to the extent that visitor needs of viewing undisturbed wildlife are excluded." Moreover, Yellowstone staff were beginning to wonder "how much human handling wild animals can endure and still retain their wild characteristics." Rather than issuing celebratory press releases about the use of immobilizing drugs and "modern devices" in the park, as Garrison had done in the first years of the Craigheads' research, the Park Service now sought to promote a gentler image of park wildlife research. In the summer of 1966, a few months before resigning from his position as the Park Service's chief scientist, Sprugel warned a researcher who had written an article titled "Yellowstone Black Bears Wear Ribbons" that some of the draft's language might provoke undesirable reactions from "relatively uninformed 'ardent animal lovers.'" One point that could be brought out more clearly, he suggested, "is that the bears are not really being 'tortured' during the immobilizing or trapping procedures. Maybe I am a little sensitive on that point but I can just visualize the letters coming in protesting the cruelty imposed upon black bears in Yellowstone by Colorado State University students with the sanction of National Park Service personnel." Even sport hunters began to express concern about the new techniques of wildlife biology. In February 1967, the "Wood, Field and Stream" columnist for the *New York Times* complained about the effect the "priests of technology" were having on wildlife: "A hunter likes to think of the game he has shot as authentically wild, untouched by human hands. But too often it's been manhandled. . . . I'm waiting for the day the porpoises capture a scientist, put ear tags in him, hang a transmitter around his neck and send him back from where he came."[33]

In early 1967, the Craigheads' research came under direct attack from the Wyoming-based nature filmmaker Walter Berlet, who began criticizing their use of radio collars in the public lectures and film screenings he gave to promote wilderness protection. John Craighead first heard about Berlet's comments from Peter Arnold, who had heard Berlet tell his audience that while the Craigheads "would have you believe that this program of putting radios around the necks of the grizzlies is to save them, actually this will seal their doom." In addition to informing Craighead, Arnold wrote to the executive director of the Massachusetts Audubon Society, for whom he believed Berlet to have been fundraising: "If a few bears had to suffer the indignity of wearing a collar with a small radio attached so that the entire species could be saved, then this is a small price to pay indeed. If the idea of using our technological skills to save wildlife is so horrendous, then we should give up bird banding, fish tagging and all other wildlife management programs." Berlet's attacks concerned John Craighead enough that he not only wrote to Berlet to ask him to cease and desist but also pleaded with Park Service leadership and senior figures associated with the National Audubon Society to help him silence Berlet. None of them proved willing to intervene, and Berlet was unwilling to back down of his own accord.[34]

In his response to Craighead, Berlet admitted that he was not a professional wildlife biologist, but he noted Olaus Murie had shared his opposition to tagging when a similar study had been proposed for McKinley's bears. He was willing to change his views, he told Craighead, "if you can show that you gained useful information that will preserve this Yellowstone bear and that this population was not damaged or endangered by your research methods." Faced with Berlet's intransigence, Craighead accused him of causing more damage to grizzlies through his speeches than radio collars ever could: "Until such time as you can meet with me for a detailed explanation of the program, I would hope that you would refrain from expressing your views in public even though you may not change them. I can assure you that your actions, which no doubt have been well intentioned, have only contributed to placing in greater jeopardy the very animal you profess to feel so concerned about."[35]

John Craighead's defensiveness in the face of Berlet's attacks was probably exacerbated by his concerns about the National Geographic television special on himself, his brother, and their research on grizzlies that was scheduled to air later that year. Though hardly unfamiliar with national media exposure, the Craigheads had almost always had authorial control over their representations, providing both text and images for their *National Geographic* magazine articles and scientific articles and books. They had far less control over the television special, which was to be broadcast with the title *Grizzly!* In the spring of 1967, John Craighead reassured the president of the University of Montana that although "we had reservations regarding publicizing our work to the extent that is necessary for a TV program," the final product appeared to be sound. A few months later he offered a similar assurance to Yellowstone superintendent McLaughlin, who had called him to express concern after seeing the script of *Grizzly!* Craighead told McLaughlin that "no one recognizes the hazards involved in presenting a film of this kind to the general public any more than Frank and I do. We did a lot of soul searching before we agreed to cooperate with the National Geographic Society in this venture." They were convinced, he added, that their research and the park's bear management policies would be responsibly represented in the final version of the script. Craighead's assurances notwithstanding, his dispute with Berlet would soon be replayed on a far more public stage and with more serious consequences.[36]

Research Itself

The broadcast of *Grizzly!* was only one of several factors that converged to bring the Craigheads' work in Yellowstone under intense scrutiny in 1967. Another was the Craigheads' completion of a major report on Yellowstone's grizzly population that included what turned out to be highly controversial management recommendations. In June 1967 Craighead sent a copy of the report to McLaughlin, whose term as Yellowstone's superintendent was ending that summer. Craighead thanked McLaughlin for "minimizing the personnel and public relations aspects of the work" in the park, adding that, "as you know, conducting research can sometimes be almost as difficult a public relations problem as it is a biological one." That month the Craigheads met with Yellowstone staff to discuss the report and their plans to continue their research in the park under grants from NSF and the AEC. The report's most controversial recommendation was that the park's open-pit dumps, including the Trout Creek dump, where the Craigheads had trapped and tagged most of their bears, should be closed gradually rather than suddenly in order to give the bears time to adapt to the loss of an important summer food source. The Craigheads agreed with the Park Service on the importance of closing the dumps, as they had since their first field season in the park in 1959, but they worried that a sudden closure would send bears into campgrounds in search of food, where some would inevitably pose a threat to humans and have to be killed. Given what they believed was the precarious status of the population, the Craigheads argued that the additional, unnecessary mortality would be unacceptable.[37]

Another important factor was the replacement of McLaughlin by a new superintendent, Jack Anderson, and the hiring of several new park biologists. Frank Craighead later described 1967 as a "watershed" year, after which the formerly "close and relatively amicable" relationship between researchers and park administration was replaced by "mistrust, suspicion, and what we could only interpret as hostility." McLaughlin's initial response to the Craigheads' management recommendations had been positive, but his opinion was not shared by the new administration, particularly Anderson and the park's new supervisory research biologist, Glen Cole, who had previously worked together in Grand Teton National Park. In the late 1950s, as an employee of the Montana Fish and Game Department, Cole had been an enthusiastic advocate for the modernization of wildlife biology in what he described as "this age of missiles and Sputniks," but in Yellowstone he and Anderson argued for strict constraints on the kinds of studies that could be conducted. In late July 1967, McLaughlin informed John Craighead that Park Service director Hartzog had "definitely put us in the research business" with the appointment of Cole and two other full-time Yellowstone biologists. The strengthening of the scientific staff made it easier for the Park Service to consider whether the benefits of having the Craigheads working in Yellowstone outweighed the costs—particularly the cost of having nationally prominent researchers use their work in the park as the basis of management recommendations that contradicted the policies of the Park Service.[38]

The new Yellowstone administration moved quickly to counter the Craigheads' effort to gain support for their position on the dump closure in the press and in the conservation community. In August 1967, Anderson told Hartzog that the Craigheads would be allowed to continue their research in the park, but only under Cole's "strict surveillance and guidance." Meanwhile, Cole prepared a strategy for the park administration to undermine the credibility of the Craigheads' management recommendations. Like Prasil, who had urged Adolph Murie to publish his findings on McKinley's grizzlies several years earlier, Cole recognized the importance of being able to call on sources of scientific expertise within the Park Service to refute the claims of outside researchers. He told Anderson that the Park Service should defend its approach to bear management by referring to extensive research on grizzlies since the 1940s by Murie, the "internationally known Park Biologist." Murie's work showed that Park Service policies had "evolved as a result of careful studies by Park biologists and the experiences of Rangers over a long period of time." Although Murie was reluctant to speak out against the Craigheads in the press, he actively supported Anderson and Cole's effort to minimize the use of research "gadgetry" in the park, later giving Cole advice privately on how to respond to the Craigheads' attacks on park policy.[39]

Murie also continued to oppose wildlife tagging in McKinley, even when the creatures to be tagged were neither threatened nor iconic of wilderness. When a graduate student in the Yale School of Forestry sought his blessing for a study that would have involved tagging McKinley's abundant arctic ground squirrels, Murie's response was just as impassioned as it had been to Dean's earlier proposals for grizzlies. Murie told the student that he would have been happy to endorse the project "if it did not involve *tagging animals* in a *national park*." He had tagged hundreds of rodents over the course of his own career, he admitted; it was the use of such methods in national parks, not their use in general, that offended him. The student's proposed tagging, "though less obvious than the flamboyant gadgetizing of larger animals, nevertheless would inflict the same type of blemish to the wilderness spirit, and would serve as an entering wedge for this type of research within the park." Murie noted that he and others had opposed other research projects in the park besides wildlife tagging, notably the proposal for an astronomical research station near the peak of Mount McKinley. Even if the station had been completely hidden from view, Murie argued, the "*mere knowledge* that such a station existed would not only have been a blight on the mountain, but would have contaminated the wilderness spirit of the entire park." If such uses of McKinley were allowed to proliferate, they would "cheapen the esthetics of this wilderness park" and transform its "charm and inspirational qualities" into a "mundane laboratory atmosphere." Murie concluded by noting that "tagging research is becoming a real threat in McKinley," but, at the same time, "the opposition to this type of despoliation in parks is increasing."[40]

On November 1, 1967, the prime-time broadcast of *Grizzly!* on CBS brought the Craigheads' research methods and their growing conflict with the Yellowstone administration to national attention. While it is hard to know exactly what effect the broadcast had on its viewers, it did nothing to simplify the relationship between the Park Service and the Craigheads. The president of the National Geographic Society told John Craighead that the film had made a positive impression on Hartzog and other Park Service staff when screened in Washington a few months earlier, but he was writing largely to reassure Craighead that a National Geographic press release about the documentary that mentioned the Craigheads' criticism of Yellowstone's grizzly management policies had not further damaged their relationship with the Park Service. A reviewer of *Grizzly!* for the *Washington Post* criticized only the title, which "led many to expect a program of violence, combat and gore," when instead the program offered "superb color photography and a well told story" of two "scientists of the wilderness" and the animals they studied and tried to protect. Others were less impressed by the program. The writer, filmmaker, and conservationist Lois Crisler wrote to Adolph Murie to ask him whether he shared her feelings about the show, which struck her as sensationalistic and disturbing in its depiction of the Craigheads' hands-on research methods. "That bear panting violently—was it killed? The cub 'restored to its mother'—just how? Walking with a receiver held out in the hand—how far? to what avail? All information given had already been found by you," she told Murie, "without drugs or markers."[41]

Crisler would have been happy to learn that the Yellowstone administration had already begun restricting the Craigheads' handling and tagging of park wildlife by the time *Grizzly!* was aired. In October 1967, in regard to the Craigheads' request to kill two elk to use as bait for grizzlies they hoped to collar, Cole told Yellowstone superintendent Anderson that he believed some restraint in the use of hands-on research was necessary "if research itself is not to become a contributing factor to the loss of park integrity as a natural area." Cole ultimately denied the request. Although the Craigheads would later accuse the Yellowstone administration, not without justification, of singling them out for special restrictions, Anderson and Cole's opposition to hands-on research extended beyond the Craigheads to other outside researchers and to Park Service biologists. The next spring, for example, when a group of scientists suggested establishing a research station just outside the park borders in the town of West Yellowstone, Anderson urged them to coordinate with the park administration so that outside researchers would not "unknowingly conduct studies that detract from the integrity of natural ecosystems" by "using park wildlife as experimental animals to test telemetry equipment or immobilization drug dosages." A few years later Anderson warned a graduate student at the University of Calgary who had proposed conducting a study of pronghorn antelope in the park that "while national parks offer unique opportunities to study natural phenomena in a relatively undisturbed state, there are . . . sometimes restrictions on study methods."[42]

After Sprugel stepped down as the Park Service's chief scientist in 1966, he was eventually replaced by Leopold, who served on a part-time basis for only a year. When he resigned in the summer of 1968, Leopold volunteered to mediate the increasingly "hostile situation" that had evolved between the Craigheads and the Yellowstone administration, which he believed was beginning to tarnish the Park Service's reputation in the national media and in the scientific community. On the basis of complaints from the Craigheads, the Yellowstone administration had received inquiries from the AEC, funder of a large portion of the Craigheads' work in the park, and from Sprugel, now the head of the Illinois Natural History Survey, about its supposed efforts to force non–Park Service scientists out of the park. At Leopold's own recommendation, Hartzog appointed him chairman of an ad hoc advisory committee on grizzly management in Yellowstone that would, they hoped, resolve the controversy once and for all.[43]

While Leopold's committee deliberated, the Craigheads continued to wrestle with the Yellowstone administration over the use of tags and collars. Despite growing tension, the Craigheads continued to plan for a long-term presence in the park, which had now been the focus of their research for nearly a decade. In January 1969, John Craighead told Anderson that he expected the AEC-funded "big game" study would probably take four or five more years to complete. The Craigheads' persistent use of terms such as "outdoor laboratory" and "big game," with their connotations of experimental manipulation and sport hunting, made it easy for critics to accuse them of failing to understand the unique mission of the national parks. Their choice of highly visible research subjects could make them seem oblivious to public opinion. Cole told Leopold he was worried about visitors' reactions to the Craigheads' collaring of elk in Yellowstone's Firehole and Gibbon areas, the location of "our main roadside scenic herd. Neckbands and radio transmitters used to date are very conspicuous and animals so marked are seen and photographed by large numbers of visitors." A public backlash, Cole warned, might prevent park biologists from tagging wild-life even when the technique was essential to park management.[44]

As the centennial celebration of Yellowstone's founding in 1872 approached, Anderson and Cole began pressuring the Craigheads not only to stop collaring additional animals but also to begin removing existing collars and tags. In April 1969, Anderson told Craighead that the "conspicuous marking of park wildlife seems to have reached the point where it detracts from the scenic and esthetic values obtained from viewing wildlife." In advance of the centennial, Anderson asked Craighead to cooperate in "removing conspicuous markers from bears to the greatest extent possible." The Yellowstone administration realized, he added, that "our wildlife which is relatively unafraid of humans affords an unusual op-portunity to immobilize and mark animals at will, but these are also animals that will be seen and photographed by thousands of visitors each year," and Yel-lowstone's own research biologists had been placed under the same restrictions that applied to the Craigheads. Perhaps, Anderson suggested, the Craigheads could pursue such research in areas managed by the Fish and Wildlife Service, such as the National Elk Refuge in Jackson Hole. Anderson and Cole's efforts to reduce tagging in the park were supported by Park Service leadership. The director of the Park Service's Midwest Region, for example, told Anderson that he agreed with the decision "to stop further conspicuous marking of wildlife" and the use of "distractive markers," since "radiotracking and biotelemetry stud-ies . . . can be fully facilitated outside of National Parks, where the presence of marked animals and the disturbance of natural conditions may not necessarily be detrimental to the study area's purpose."[45]

In the spring of 1969, the Craigheads reluctantly signed a new memorandum of agreement with the Park Service stating that certain scientific practices were not compatible with the Yellowstone's status as a natural area. They also agreed to cooperate in removing conspicuous markers and to end the grizzly study within two years, though John Craighead told Anderson he hoped that cer-tain new studies involving satellite tracking of wildlife could still be conducted, provided "the time, place, and number of animals involved would not conflict with centennial objectives." Craighead noted that, in his opinion, the aesthetic impact of tagging was much less than Anderson seemed to believe, and the ben-efits far greater, as many wildlife biologists and park administrators had recog-nized: "At the risk of appearing immodest, I think I can say that the techniques of color marking, immobilizing, handling, radiotracking, and data gathering that we, our colleagues, and our students developed or perfected in the course of ten years of research effort in the Park are now being widely applied in other national parks throughout the world."[46]

In the fall of 1969, Leopold's ad hoc advisory committee on grizzly bear management in Yellowstone issued its report. Unsurprisingly, given Leopold's close association with Park Service leadership, the report recommended the immediate closure of the dumps and firm limits on scientific practice within the park. The report did not, however, have the effect that Leopold and Hartzog had hoped it would have. A month after it was issued, Cole was still frustrated by the Craigheads' success at gaining the ear of the national media, writing to Leopold to confess that "this kind of biopolitics is beyond me."[47]

Cole was more adept at the kind of biopolitics that could be practiced within the borders of Yellowstone. The Craigheads' heavy dependence on tagging made them vulnerable to interference by park staff, a fact the Yellowstone administration quickly recognized. Frank Craighead later described the decision in 1969 to prohibit further marking or radio tagging and to require the removal of existing tags as the single change in park policy that "would damage our study more than any other." By the time the Craigheads' research in Yellowstone came to an end, they had marked a total of 264 bears. Frank Craighead later explained that "visual marking had served from the beginning as our principal method for positively identifying bears; our ability to systematically gather reliable data on grizzly movement, population, and mortality depended on it." The Park Service's removal of the tags in the name of wilderness values and park aesthetics directly undermined the Craigheads' ability to acquire knowledge that could be used to challenge the Park Service's new policies.[48]

Despite these constraints and their disintegrating relationship with the Yellowstone administration, the Craigheads still hoped to use park wildlife for the development of satellite-based wildlife tracking techniques. This was a possibility they had been pursuing since a 1966 conference at the Smithsonian Institution organized by Sidney Galler, who had left ONR to become the Smithsonian's assistant secretary for science in 1965. In collaboration with Helmut K. Buechner, a wildlife biologist at the Smithsonian, and Charles E. Cote, an engineer at NASA's Goddard Space Flight Center in Maryland, the Craigheads had proposed using NASA's experimental Interrogation, Recording, and Location System to monitor the seasonal migration of Yellowstone's elk. Although IRLS was far from ideal for wildlife tracking, they hoped that a successful demonstration using the system would help convince NASA to launch a more suitable instrument.[49]

Yellowstone was the Craigheads' first choice as a site for attaching the satellite collar because of their familiarity with the park and the tameness of the elk. Neither Anderson nor the superintendent of nearby Grand Teton National Park, however, would grant permission for the study. Buechner had convinced S. Dillon Ripley, the secretary of the Smithsonian, to appeal to Russell E. Train, the undersecretary of the Department of the Interior, to overrule the superintendents' decision, but Train deferred to the Park Service's Natural Science Advisory Committee, chaired by Leopold, which sided with the superintendents. Still, Train instructed the Park Service to cooperate with the researchers if the elk migrated through Teton or Yellowstone as expected. By this time the Craigheads' conflict with the Yellowstone administration had gotten so much attention in the press that one University of Colorado graduate student told Anderson that he was planning to use it as the subject of his thesis on "the role of scientific information in the administrative decision-making process."[50]

Late in the fall of 1969, despite the general prohibition on tagging and the Park Service's refusal to allow an elk to be collared for the satellite-tracking experiment, the Craigheads were granted permission to radio-collar one of the park's black bears, who they then tracked to his winter den. In February they returned to instrument the bear with physiological sensors and install communications equipment that would allow them to be monitored by satellite. They also tested a satellite collar on an elk in the National Elk Refuge in Jackson Hole after having previously tested a mock-up of the collar on a captive elk. Although the collar was bulky and weighed nearly 25 pounds, it seemed to have no untoward effects on the captive elk. After releasing her back into the herd, the researchers hoped to recapture her to attach the working collar. On the day of the attempt, however, she remained skittish and out of reach of the tranquilizing dart. In the attempt to single her out in the dense herd, another female elk was accidentally struck. Since this new elk appeared healthy, the Craigheads decided to collar her instead, dubbing her "Monique." However, a malfunction in the collar almost immediately prevented it from communicating with the satellite, and the elk herself sickened and died on February 25, 1970, just six days into what the researchers had hoped would be a several-month-long study.[51]

The Craigheads soon regretted having invited the press to the collaring to drum up enthusiasm for wildlife satellite tracking. The widespread media coverage of Monique's haphazard collaring and subsequent death soon resulted in hundreds of angry letters to the researchers and their sponsors. In the spring 1970 issue of the newsletter of the Defenders of Wildlife, editor in chief Mary

Hazel Harris included a photograph of the collared elk apparently struggling to stand, under the headline "'Monique' Death Ends Project." Many readers followed her recommendation to express their disapproval in writing to NASA and the Smithsonian. The newsletter of the Animal Welfare Institute reprinted a letter from one professor of entomology who had criticized the elk study in his local newspaper in central Illinois as an "awkward, cruel, and nonsensical 'experiment,'" a "glorified high school Science Fair project," and a "grandstand play for publicity." In response to these criticisms, Ripley, the head of the Smithsonian, was forced to defend the project to members of Congress who had heard about Monique from their constituents, while the president of the Wildlife Society asked Buechner for help in responding to the numerous telephone calls he had received relating to "the public's reaction to the transmitter elk which recently died (?) out west." Letters to the Smithsonian continued to trickle in for years after the event, attacking the collaring as cruel, wasteful of taxpayer money, and equally irrelevant to science and conservation.[52]

Although the Craigheads had more success satellite-tracking a second elk later that spring, the Monique fiasco seriously undermined NASA's support for the technique. In June 1969, NASA convened a meeting on wildlife satellite tracking in Washington that was attended by Warner, who had first proposed the technique with Cochran in 1962, as well as the head of the Fish and Wildlife Service and the collaborators on the Monique project. At the end of the meeting, NASA deputy administrator George Low told the assembled researchers and engineers that "if it's worth doing, it's worth doing well. And the results to date have—I think it's been less than we'd like to see in some other programs." Though NASA did not entirely cut off support for wildlife satellite tracking, it did stop directly supporting fieldwork, most likely slowing the development of the technique. Looking back on the Monique incident a few years later, Buechner blamed NASA's mishandling of the publicity for the extent of the public protest. Attempting to reassure a research collaborator who was concerned about the potential for similar problems in a proposed satellite-tracking study of African elephants, Buechner explained that the letters he had received "were amazingly stereotyped, indicating a response from a certain type of person in society."[53]

The public response to Monique's death strengthened the Park Service's conviction that such research was best performed outside the parks. At the same time, the Yellowstone administration worried about providing fodder for accusations that it was suppressing scientific research in the park. In the summer of

An artist's rendition of satellite-tracking of an elk was distributed in a press kit in early 1970 by Radiation, Inc., the company that manufactured the collar in collaboration with the Craigheads, Buechner, and Cote. (Courtesy of the Smithsonian Institution Archives)

1970, the Craigheads again proposed conducting a satellite-tracking experiment in Yellowstone. Leopold advised the Park Service against approving the Craigheads' proposal, mainly because they had proven so difficult to work with in the past, but Anderson noted that the administration was in "a rather delicate position because we are, in fact, promoting research in the park." A few weeks later

Anderson told John Craighead his request had been denied and reiterated his desire to have most of the park's conspicuous wildlife tags removed before the centennial in 1972, when the park would host scientists, conservationists, and park administrators from around the world. A clearly exasperated Craighead replied that banning the technique would only retard the progress of science in what he called "one of the finest outdoor laboratories for ecological research that exists anywhere in the world." He denied that his team had been or would be "responsible for large-scale, conspicuous marking of wildlife in Yellowstone in areas of intensive visitor use"; on the contrary, they expected to have "at most three or four animals marked or instrumented at any given time." While there were many places outside Yellowstone where biotelemetry research could be conducted, he noted, few contained populations of "big game" that were as tame as those within the park. Finally, the data that radio tracking provided was so useful that Craighead predicted "that if you are going to learn more about grizzly bear–man relationships in and around campgrounds your biologists will very soon themselves resort to the use of radios."[54]

In 1970, Anderson's antitagging policy was so strictly applied that seven "problem bears" were transplanted by park rangers without being given ear tags. Robert Linn, who had succeeded Leopold as the Park Service's chief scientist, supported Anderson's move to reduce wildlife tagging in Yellowstone as a way of improving the visitor experience. In January 1971, defending the new policy to the president of the Wildlife Society against accusations that it was aimed solely at driving the Craigheads out of the park, Linn noted that Park Service biologists had also been "guilty" of excessive tagging and would be subject to the same restrictions:

> Research projects in national parks, regardless of whether such projects are mission-oriented or disinterested research, must be designed so that the project will neither have an adverse impact on the park resources nor on visitor enjoyment of those resources. In this regard, we recognize the value of tagging and otherwise marking animals for identification purposes, but when the majority of bears, as well as many of the elk, bison, and moose the park visitor sees are wearing radios, tags and brightly colored streamers as they were in Yellowstone, we are of the belief that the visitor's opportunity to enjoy park wildlife has been seriously impaired.[55]

By 1971, Yellowstone staff had resumed ear-tagging bears that were translocated, recognizing that data about the bears were essential to defending the park's management approach against its numerous critics, including the Craig-heads. In March 1971, one park resource management specialist emphasized the importance of ear-tagging to Yellowstone's rangers in light of the recent closure of all the park's dumps: "Our bear management program may be severely tested this year, and while ear tags are undesirable from the standpoint of wildlife viewing, the positive identification of problem bears is critical to the program's success. *All* grizzlies transplanted including cubs are to be eartagged before release."[56]

Though recognizing the usefulness of inconspicuous tags on problem bears, the Yellowstone administration continued to remove the Craigheads' more visible tags. In the summer of 1971, John Craighead learned that some of his team's tags had been removed by park rangers, apparently at Anderson's request. He wrote to Anderson to protest what he saw as an act of scientific sabotage, pointing out that, in his estimation, only 1 percent of visitors ever encountered a marked bear. Anderson replied that park rangers had misunderstood his instructions. Only the Craigheads' brightly colored plastic ear tassels, he explained, were to be removed in advance of the centennial celebration; the smaller ear tags would be allowed to remain. He assured Craighead that he and his brother had been asked to terminate their study because of their disparagement of the Park Service in the popular press and their refusal to share research findings, not because of their research methods. But he also stressed the importance of finding an alternative to "the conspicuous marking of park wildlife that is seen and photographed by Yellowstone visitors" and pointed out that 1 percent of Yellowstone's annual visitors still amounted to 25,000 people who were "short-changed in seeing the grizzly as it occurs naturally."[57]

In the fall of 1971, the Craigheads refused to sign a new memorandum of agreement that would have given the Park Service firm control over their research methods and public statements, thereby bringing to an end nearly twelve years of intensive research in the park. Even after the Craigheads had left, Park Service leadership continued to assert the need to minimize handling of the park's animals. That fall Hartzog informed the director of the Bureau of Sport Fisheries and Wildlife that capture and immobilization would no longer be used for research in Yellowstone. The only bears that would be captured or handled would be "problem animals that must be removed from camp grounds and other visitor concentration areas." Ear tags would be left on bears that had previously been marked, and new ear tags would be attached to unmarked bears, but none of the more conspicuous tagging methods would be allowed. A year later Linn explained to Montana senator Mike Mansfield that "the less the grizzlies are handled and manipulated the better, and the closer the day that grizzlies in Yellowstone will be wild and free-ranging once more."[58]

The Last of the Old-Time Field Biologists

The closure of the dumps and the termination of the Craigheads' research in Yellowstone did not end the debate over grizzly research and management. On the contrary, the controversy intensified. In 1973, in recognition of the fact that neither Park Service biologists nor the Craigheads were impartial observers of the bear population, a new Interagency Grizzly Bear Study Team was assembled to coordinate grizzly research in the Greater Yellowstone area. From its inception, the IGBST fought for the right to tag and radio-collar bears. In the summer of 1973, one member of the team, Robert B. Finley of the Denver Wildlife Research Center, urged Anderson to halt the removal of the Craigheads' color markers. The "firm data on bear movements" that they could provide, Finley argued, outweighed the "objections of photographers and the public." Anderson replied that the markers were being removed because they were damaged and of dubious scientific value, not because of their effect on the visitor experience, but he reiterated his belief that alternatives to "unsightly streamers" should be sought. More than a year later—and three full years after the Craigheads had left Yellowstone—Anderson again argued against the use of tags. Don Frickie, the manager of the Arctic National Wildlife Refuge, had written to ask him if Yellowstone's rangers and naturalists had observed "personality change" in bears that had recently been drugged and tagged. Anderson told Frickie that such concerns had contributed to Yellowstone's decision to stop marking bears "unless absolutely necessary to determine if we had in fact handled a bear previously." Though Anderson's attacks on the Craigheads' research methods were often dismissed, at the time and in retrospective accounts, as secondary to personality conflicts or policy disputes, his opposition to hands-on techniques was persistent and principled.[59]

Anderson's personal opposition to tagging was not enough to allow the Yellowstone administration to resist the enormous pressure that was soon brought to bear in favor of the technique. One source of pressure was a new report by an advisory committee led by the prominent Canadian wildlife biologist Ian McTaggart Cowan, who had previously played little role in the controversy. Cowan's committee harshly criticized the studies on grizzlies conducted by the

Park Service since 1970 and urged that independent research be strengthened. Another source of pressure was Park Service and Department of Interior leadership in Washington, who were eager to put the grizzly controversy behind them, even if that meant capitulating on the tagging issue. In late 1974, in response to an inquiry about the Craigheads' departure from Yellowstone, Nathaniel Reed, Train's successor at Interior, emphasized that "it has [been] and will continue to be our policy to allow research so long as it is not detrimental to the ecosystem."[60]

Reed's emphasis on damage to the Yellowstone "ecosystem" indicated an important shift in the guiding principles of park management from the aesthetic concerns with wilderness experience that had motivated the Muries to new concerns with ecosystem function and endangered species protection. As Biel has argued, the new image of Yellowstone's bears that emerged in the 1970s was "the imperiled bear, hanging on to survival by the tips of its claws"—an image given legal force by the listing of the northern Rockies grizzly bear population as "threatened" under the Endangered Species Act in 1975. Richard Knight, head of the IGBST, later claimed that he had to "use psychology" on Anderson and Cole in order to get approval for radio tagging during the first few years of the IGBST by telling them that the technique would reveal things the Craigheads had known nothing about, but the effectiveness of Knight's manipulation depended on these broader structural shifts, which pushed aesthetic concerns into the background. In late 1975, when Knight asked Cole for permission to replace the collars on two grizzlies who were denning inside Yellowstone, Cole granted it without delay. By 1977 a large-scale trapping and tagging effort was under way; within a decade, the IGBST had collared ninety-seven bears in and around Yellowstone.[61]

Radio tagging also made its way into the wilderness stronghold of McKinley. The long-range wildlife management plan for the park that had been drawn up in 1970 had stipulated that any scientific research would be "in keeping with basic Park policy. The premise that the end justifies the means will always bear close scrutiny." Restrictions on tagging meant that when Dean and his graduate students began a study of the impact of road traffic on wildlife behavior in McKinley in 1973, they were limited to the same strictly observational techniques that Dean and Murie had used in the late 1950s. In 1976, however, the McKinley park administration began considering whether to allow a radio-collaring study of the park's caribou, whose numbers had recently suffered a dramatic decline. If the decline was due to natural causes, the Park Service would allow natural processes to continue working; if it was due to human causes, it would intervene. The radio-tracking study, it was argued, would help determine which was the case; it was a small intervention to determine whether a larger intervention was justified.[62]

The caribou study was to be carried out by Troyer, whose pioneering efforts to trap and drug brown bears on Kodiak Island in the late 1950s had helped inspire Dean's proposals to radio-tag McKinley's grizzlies. In contrast to the Craigheads' or Dean's proposals a decade and a half earlier, Troyer's proposal to radio-collar McKinley's caribou was subject to a public consultation process. In February 1976, as head of the Biology and Research Program of the Alaska Cooperative Park Studies Unit, Dean attended a meeting of state and federal agencies in Fairbanks at which Troyer argued that the aesthetic impact of the study would be minimal. A month later, Troyer presented the study proposal to the members of the Denali Citizens Council, an organization based in the small commercial area just outside the main McKinley park entrance. The council concluded, in the words of its chairman, Tom Adams, that the small number of animals involved in the study, their distance from areas frequented by park visitors, and the fact that the radio collars were "relatively unobtrusive" diminished "the likelihood of visitors being offended by seeing these marked animals." In light of the herd's rapid decline, the information produced by radio collaring seemed worth the costs. An environmental impact statement completed by the Park Service a few months later identified the main drawbacks to the study as handling a previously unhandled caribou herd and affecting the aesthetics of the park, but it too concluded that these costs were well justified by the possibility of understanding why the park's caribou were no longer thriving.[63]

In 1981, the scientific monograph on McKinley's grizzlies that Murie had told Prasil was nearing completion in 1965 but which he continued revising until his death in 1974 was published posthumously by the Park Service. In his preface to the book, Adolph Murie's son Jan Murie noted that his father had held "strong philosophical views about biological studies in national parks. . . . Although he recognized that studies of marked bears would yield additional data of value, he felt strongly that marked animals are out of place in national parks." The last chapter of the book, "Keeping Grizzlies Wild," featured arguments against tagging and other forms of hands-on research taken directly from the letters Murie had written in the early 1960s in opposition to Dean's proposals for McKinley. The book concluded with a plea for idealism that linked the wildness of the grizzlies to the state of American culture: "The national park idea is one of the bright spots in our culture. The idealism in the park concept has made every American visiting the national parks feel just a little more worthy. Our generosity to all creatures in the national parks, this reverence for life, is a basic tradition, fundamental to the survival of park idealism. Perpetuation of truly wild grizzlies in McKinley National Park is essential to maintain this tradition."[64]

The biologists who reviewed *The Grizzlies of Mount McKinley* were divided on the value of Murie's observational methods and anecdotal reporting, but they all saw Murie as a voice from a lost age. Wildlife biologist Robert Rausch, who was sympathetic to Murie's concerns about hands-on research, found the amount of information in the monograph "remarkable, all the more so because Murie relied solely on the observation of free-ranging animals without use of capture-guns, marking of animals, radiotelemetry, and similar methods that seem to have become *conditio sine qua non* in most contemporary studies of large mammals." Bear biologist Charles Jonkel, in contrast, "grimace[d]" at Murie's failure to quantify and rigorously evaluate the rich observations he had obtained, though he also admitted a kind of nostalgia for the research methods and values that Murie represented: "The spirited cry for allowing the grizzly dignity and freedom in his domain, rather than intensive, modern-day research and management programs, is like a last glimpse of a beautiful world gone forever."[65]

Though the Craigheads could no longer study Yellowstone's bears firsthand, they continued to participate in debates over Park Service policy. In 1979 the Sierra Club Press had published Frank Craighead's *Track of the Grizzly*, which included a chapter titled "Bureaucracy and Bear" that accused Anderson and the Park Service of risking the survival of the grizzly for the sake of bureaucratic turf. In 1982, John Craighead and several of his research collaborators published *Definitive System for Analysis of Grizzly Bear Habitat and Other Wilderness Resources*, a report on their use of Landsat satellite imaging to map current and potential grizzly habitat in the northern Rockies. In his preface to the report, Hornocker, who had been the first of Craighead's graduate students to work on Yellowstone's grizzlies, argued that the hands-off methods it described were, though groundbreaking, insufficient on their own. Satellite images could be used to map grizzly habitat only "because the grizzly bear's biological parameters and biological needs had been documented earlier" through techniques such as "capture and drugging, individual color-marking, radio-instrumentation and remote-sensing of animals." Satellite imaging, in other words, was a complement to, rather than a replacement for, techniques that required handling of wild animals. While there were "those who consider the grizzly bear to be a sacred wilderness animal not to be captured, handled or marked for scientific study," Hornocker wrote, the observational techniques advocated by these critics would not suffice at a time when even federally designated "pristine" wilderness areas were under threat by a new antienvironmentalism. However useful satellite imaging might be, Hornocker concluded, wildlife biologists would have to continue trapping, drugging, and marking the animals they wanted to save.[66]

In contrast to Hornocker's uncompromising preface, the body of the report suggested that the IGBST's plans for extensive radio collaring of the Greater Yellowstone population were excessive, unnecessary, and likely to stress the population. By the early 1980s, John Craighead seems to have become troubled by the rampant proliferation of radio-tagging studies that, unlike his and his brother's work in Yellowstone, were unjustified by any clear scientific goals. In 1982, in the entry on grizzly bears for a volume titled *Wild Animals of North America: Biology, Management, and Economics*, he and a colleague argued that artificial marking was necessary for obtaining accurate biological data about bears. After providing details about the use of drugs, tags, and radio collars, however, they also noted what they saw as "justifiable concern that agencies may have over-responded to the plight of a threatened species with a surfeit of research." Some continued monitoring of threatened populations was necessary, they argued, but intensive studies to obtain basic biological parameters did not need to be repeated with each and every population of bears. Because capture and tagging put stress on bear populations, they recommended keeping handling to a minimum; it could "hardly be justified when used as a continuous monitoring and data-gathering technique in the ongoing management process."[67]

Craighead reiterated the point a year later in an interview with wildlife biologist Les Pengelly that was published in *Defenders*, the magazine of the Defenders of Wildlife. The introduction to the piece described Craighead as perhaps "the last of the old-time field biologists, in the tradition of the Muries, who combine a wide-ranging knowledge of botany, geology and wildlife with remarkable physical skills." Craighead told Pengelly that large-scale marking of grizzly bears had been essential for acquiring "the same vital statistics that actuarial people get for human populations when they want to predict longevity, birth rates and population behavior," but that such research was no longer necessary or even in most cases acceptable. When Pengelly asked him what needed to be done to save Yellowstone's grizzlies from extinction, Craighead replied that "we must no longer encourage and support intensive population research. . . . We need to manage them, not research them." In 1984, after a summer character-ized by an unusual number of attacks by grizzlies on humans in the Yellowstone region, the Montana Fish and Wildlife Department began looking into whether repeated darting with muscle relaxants and anesthetics was responsible for the bears' aggressive behavior. Craighead told the press that there were many other possible explanations and warned that advocates of bear hunting might use the idea of "drug-berserk" bears as an excuse to relax needed protections, but he also argued that the issue "raises enough questions that we should use some restraint in drugging bears."[68]

In contrast to John Craighead's newfound concern about hands-on research, the Park Service in the 1980s enthusiastically embraced radio tracking of its iconic "wilderness wildlife," including grizzlies and wolves, the two species that had been of most concern to Murie. In 1986, in response to the poaching of five wolves within the area of the old Mount McKinley National Park (which had become the "wilderness core" of an expanded and renamed Denali National Park and Preserve) and to concerns about the legal killing of wolves in the sur-rounding preserve, the Park Service sponsored an intensive radio-tracking study of the population. The study was led by David Mech, who had been capturing and collaring wolves since the late 1960s. Between 1986 and 1994, Mech and his colleagues radio-collared 147 wolves from thirty different packs in the Denali area, some of them multiple times.[69]

Like the Craigheads, Mech looked to the Muries as a model for how to study large predators, even as he embraced techniques that they had rejected. Mech had dedicated his 1970 book, *The Wolf: Ecology and Behavior of an Endangered Species*, to Adolph Murie, and his coauthored 1998 book, *The Wolves of Denali*, opened and closed with references to Murie's classic research in McKinley. As Mech explained in a chapter in the latter work titled "Technology Yields the Data," however, there were profound differences between his team's methods and those of Murie. "In areas as extensive as Denali National Park and Preserve, the only way one can study an entire population of wolves and their prey is to use the latest technologies available," Mech argued, which meant using airplanes and helicopters to locate wolves, darts and drugs to immobilize them, and radio collars to identify and relocate them. Nonetheless, Mech emphasized that the Denali ecosystem remained much as it had when Murie began his wolf research there some six decades earlier, before any of its creatures "sported radios." The major difference was that the lives of its wildlife "were much more of a mystery then than now." A few years after Mech's wolf study had begun, Park Service biologists also began immobilizing and radio-collaring Denali's grizzly bears.[70]

Even as the Park Service embraced radio tracking on an institutional level, many members of the service shared the concerns that Murie had first raised several decades earlier. The first issue of the newsletter *Yellowstone Science*, published in 1992, featured an article on the ethics and aesthetics of radio tagging. In 2002, the Park Service's Biological Resources Management Division and Soundscapes Program Center invited Mech to conduct a review of radio tracking in the national parks. The resulting report, which Mech coauthored with Shannon Barber, noted that "some NPS staff are concerned about actual or potential intrusiveness of radio-tracking. Ideally, wildlife studies would still be done but with no intrusion on animals or conflict with park visitors." But after reviewing the advantages and disadvantages of radio tracking and comparing it to other ways of learning about park wildlife, the report concluded that the technique was often the only way of gathering knowledge necessary for management. It recommended that the Park Service conduct formal surveys to determine park visitors' attitudes toward the technique, educate visitors about its benefits, and promote the use of the least intrusive forms of tracking technology, such as satellite and GPS tags that could reduce the amount of time spent by researchers in the field. A few years after Mech and Barber's report was published, Biel conducted the kind of survey of attitudes toward wildlife tagging that it had recommended. Although she reported that "the debate over whether wild animals living in national parks and wilderness areas should be collared for scientific monitoring purposes has continued to rage," she found that only a very small proportion of visitors to Yellowstone expressed any concern about the technique; some even told her that viewing collared wildlife enhanced their enjoyment of the park.[71]

By the early twenty-first century, Adolph Murie's contributions to park science and predator protection were celebrated by wildlife biologists, conservationists, and the Park Service, but his criticisms of tagging and other forms of hands-on research had been largely forgotten. In August 2004, a new Center for Science and Learning located at the main entrance to Denali National Park and Preserve was dedicated to the memory of Adolph Murie and other members of the Murie family, whose work had contributed to preserving what was described as the "biological integrity" of the national parks. In the Murie Center's interpretive materials and in those of the nearby Denali Visitor Center,

Murie effectively became Denali's patron saint of wildlife research, including those forms of research that Murie had believed would destroy the area's "wilderness character." One article by a member of the Murie Center staff published in 2005 described the center's use of high technology for education and research, including plans for a wireless network that would extend along forty miles of the park highway and make it possible "to communicate from the field . . . to anywhere in the world via the Internet." The article also noted that, with support from ONR, the park was developing new kinds of radio collars that would make it possible "to track wildlife 24 hours per day, 7 days a week." Echoing the homage to Adolph Murie's pioneering research that Mech had included in *The Wolves of Denali*, an exhibit in the lobby of the Murie Center identified Murie as the forefather of modern-day radio-tracking studies of Denali's wolves. Perhaps needless to say, his decades-long campaign against wildlife tagging in the parks went unmentioned.[72]

CHAPTER THREE

Diplomatic and Political Subtleties

In the 1960s, Indian biologists such as E. P. Gee and Sálim Ali began to warn that the country's population of tigers would soon be extinct if habitat loss and hunting continued at their present rates. By the end of the decade they were joined by a number of European and American conservationists, including S. Dillon Ripley, the secretary of the Smithsonian Institution, who had previously collaborated closely with Ali on studies of South Asian birds. In a speech at the Bombay Natural History Society in 1967, Ripley stated that he believed tigers would be extinct in the wild within twenty-five years unless immediate action was taken to protect them. As historian Michael Lewis has argued, Indian biologists hoped that such statements from prominent foreigners would help strengthen their position with the Indian government. In 1969, Ripley repeated his warning at a meeting of the International Union for the Conservation of Nature in New Delhi, where he also promised the assistance of the Smithsonian to India and any other nation that sought to protect its remaining tigers. A census of India's tigers completed in the following year by the Indian conservationist Kailash Sankhala suggested that only 2,500 tigers remained in the country, a dramatic decline from an estimated 40,000 tigers at the beginning of the century.[1]

After the New Delhi meeting, the Smithsonian struggled to make good on Ripley's promise of technical assistance. While some Indian scientists welcomed the Smithsonian's potential contributions, others, particularly those affiliated with the Indian Forest Service such as Sankhala, resented what they saw as foreign efforts to gain influence over conservation in India. The Indian Forest Service's suspicions were heightened in the case of the Smithsonian by the latter's recently revealed entanglement with covert American military and intelligence operations. In 1969 the Smithsonian had become embroiled in a controversy over its participation in the Pacific Ocean Biological Survey Program, which critics alleged was a component of the U.S. Army's biological weapons program. After the scandal broke, Senator J. William Fulbright warned Ripley that it would be "very wise" for the Smithsonian to avoid accepting any further funding from the Department of Defense. As Lewis notes, it did not help the Smithsonian's reputation that Ripley had headed the operations of the Office of Strategic Services, the predecessor of the Central Intelligence Agency, in Ceylon during the Second World War. Dealing with the increasing sensitivity of military-sponsored research was one of Sidney Galler's main responsibilities as the Smithsonian's assistant secretary for science from 1965 to 1970. In 1968, Galler urged Ripley to increase the amount of training that Smithsonian-funded researchers received before going abroad, reminding him that "during these troublesome times when the activities of scientists no matter how pure in intent are not above suspicion, and when indeed the non-friends of the U.S. overseas appear to be multiplying, it is doubly important that scientists representing the Smithsonian Institution abroad have a full appreciation of the diplomatic and political subtleties that may affect their projects."[2]

The Smithsonian's ability to contribute to tiger conservation was also hampered by the lack of a specialist in large carnivore ecology on its staff. To compensate for this limitation, it recruited Maurice Hornocker, the former graduate student of John Craighead, who had completed his doctorate on Idaho's mountain lions under Ian McTaggart Cowan. Hornocker had since become the head of the Idaho Cooperative Wildlife Research Unit and was eager to expand his research beyond cougars to other species of large cats. In July 1971 the Smithsonian sent Hornocker to India to assess the possibilities for tiger research; on the way there Hornocker made a brief visit to the headquarters of the International Union for the Conservation of Nature in Switzerland, which was trying to coordinate international tiger conservation efforts. Hornocker met with a number of Indian officials and scientists during his trip, though heavy monsoon rains prevented him from reaching Bombay, where he had planned to meet with members of the Bombay Natural History Society, the Smithsonian's strongest scientific supporters in India. After returning to the United States, he told the Smithsonian that he was "not enthusiastic about participating in a country-wide census" of India's tigers, which he thought could best be carried out by Indian scientists with IUCN's support, but that he would be interested in participating in an intensive ecological study of the sort he had carried out with mountain lions. To lead the fieldwork in India, Hornocker recommended John C. Seidensticker, who, like Hornocker himself, had been a research assistant for the Craigheads in Yellowstone and was now completing a doctoral thesis under Hornocker, for which he had conducted the first radio-tracking study of mountain lions. At least one Smithsonian administrator, I. E. Wallen, was concerned by Hornocker's apparent attempt to "sidetrack Dr. Ripley's promise onto *another* research area" besides the urgently needed countrywide census, but ultimately the Smithsonian decided to accept Hornocker's recommendation of an intensive behavioral and ecological study.[3]

Funding for the study posed another challenge. The Smithsonian initially planned to rely on funds available through the Public Law 480 program, a scheme for international aid under which India purchased American wheat in rupees that could then be spent by American agencies working in India. Since the mid-1960s, the Smithsonian had served as a clearinghouse for the use of PL 480 funds in India, which were mainly used to support scientific and cultural work. These virtually unlimited funds inspired grand ambitions on the part of the Smithsonian's leadership. In particular, Ripley hoped to establish three permanent research stations in India along the lines of the long-established Smithsonian Tropical Research Institute in Panama. One would focus initially on tigers at Corbett National Park, another on Asiatic lions at the Gir Forest, where the Smithsonian had already been funding research for several years, and the third on rhinoceroses at Kaziranga National Park. When the United States cut off economic aid to India in December 1971 as punishment for the country's conflict with Pakistan, however, the India government retaliated by putting all proposed PL 480 projects "under review" indefinitely.[4]

Hopeful that the troubles with PL 480 were only temporary or that another source of funding could be found, the Smithsonian continued to plan for tiger research in India. By February 1972 Seidensticker and Hornocker had drafted a preliminary proposal for a four-year study of "the dynamics of an unexploited tiger population," the results of which would be "directly applicable to other tiger populations in India and throughout much of its geographical range." The study would rely on "capturing, immobilizing with drugs, individually marking and releasing tigers for future identification and study in the field," and the use of radio tracking would allow tigers to be studied "undisturbed" throughout the year. These methods had already proved useful, they wrote, for research on solitary big cats living in a densely vegetated environment—namely, the mountain lions that both Hornocker and Seidensticker had been studying in Idaho. In March, with the proposal in hand, Seidensticker and two Smithsonian staff members traveled to New Delhi, where they met the Smithsonian's liaison to the Indian government, talked with a number of Indian officials, and toured Corbett National Park, where they hoped to carry out the study. Although Indian officials would not say anything definite about the study's likelihood of approval, Seidensticker took it as a promising sign that they were willing to talk about the practical details, including plans for darting and radio collaring.[5]

With the availability of PL 480 funds still uncertain at best, the Smithsonian began searching for alternative sources of support. A month after Seidensticker's visit to India, the World Wildlife Fund, which had been founded a little more than a decade earlier as a fundraising mechanism for IUCN, announced the launch of Operation Tiger, an effort to raise more than a million dollars for tiger conservation. Most of the money was intended to support the Indian government's new Project Tiger initiative, though some of the funds would also go to other countries with tiger populations. Although it was headquartered in Switzerland, WWF's funds were raised almost entirely by semiautonomous "national appeals." As chair of the board of WWF's U.S. Appeal and a member of IUCN's board, Ripley argued that some of the funds raised for Operation Tiger in the United States should be used to support the proposed Smithsonian project. Ripley's proposal exacerbated existing tensions between WWF-US and the WWF International office in Switzerland. Although WWF's national appeals regularly sent one-third of their income to WWF International for its unrestricted use, WWF-US instead sent money on a case-by-case basis and occasionally ignored WWF International's requests to fund or not to fund particular projects. By the late 1960s, tensions over funding and other issues had resulted in what the American aviator Charles Lindbergh, a member of WWF-US's board, described as a bureaucratic "melee" between WWF-US and WWF International. The mostly European leadership of IUCN and WWF International saw Ripley's proposal as an attempt to divert money that had been raised for urgent conservation needs, such as fencing, guards, and equipment for India's new tiger reserves, toward the Smithsonian's esoteric scientific inter-ests, as well as further evidence of WWF-US's unwillingness to accept WWF International's primacy.[6]

Despite the fact that the Smithsonian's proposal had the support of several prominent Indian scientists and conservationists, IUCN and WWF Interna-tional also accused it of "scientific imperialism." In May 1972, Frank Nicholls, IUCN's deputy director-general, scolded Zafar Futehally, a member of the Bombay Natural History Society and leader of WWF-India, for having implied in a recent article on tiger conservation that IUCN and WWF would sup-port the Smithsonian's research. The proposed "research project on tiger home range studies by the use of radio-telemetry," he told Futehally, was "a purely Smithsonian programme and IUCN has no present involvement in it." Nicholls sent copies of the letter to Ripley and to Peter Jackson, a journalist serving as the head of Operation Tiger, along with a paper on scientific imperialism recently drafted by Gerardo Budowski, IUCN's director-general. Budowski's paper iden-tified scientific imperialism as a "widespread phenomenon" that was particularly "insidious" because it "can be camouflaged under what may appear to be very good goals," including claims that "all the actions are 'carried out in the name of science.'" The paper went on to enumerate a variety of forms of scientific imperialism, including the failure to adequately acknowledge local collaborators in scientific publications and the introduction of modern traps, guns, and other technologies before local cultures were "fully prepared to use them wisely on [a] truly permanent basis for the right kind of development." When Ripley for-warded Nicholls's letter and Budowski's paper to the Smithsonian staff respon-sible for the tiger project, he included a note stating that IUCN was "crazy" to associate the Smithsonian proposal with scientific imperialism "unless they are currying favor for themselves."[7]

In the summer of 1972, Seidensticker and Smithsonian administrator Michael Huxley attended the Second World Conference on National Parks, held in Yellowstone National Park and Grand Teton National Park on the centennial of Yellowstone's founding. Their encounter with the Indian delegation there dampened their hopes of carrying out the tiger study in Corbett National Park or any other tiger reserve in India. Echoing the arguments that Anderson and Cole had recently used to put an end to the Craigheads' grizzly- and elk-tagging projects in Yellowstone, Sankhala, now the head of Project Tiger, told Seiden-sticker that he was opposed to turning Corbett into a "Smithsonian laboratory." India had no need of American scientists to protect its tigers. Seidensticker's and

Huxley's disappointment over the Indian delegation's response was soon miti-
gated, however, by the warm reception they received from a Nepalese expatriate
named Kirti Man Tamang, who was also attending the conference. After rising
through the ranks of the Nepalese forest service, Tamang had taken a job in the
late 1960s at the Tiger Tops tourist lodge in Chitwan, a former royal hunting
preserve. Located in the Terai, the lowland area on Nepal's southern border
with India, Chitwan (or Chitawan) contained one of the country's few remaining populations of tigers and had just been declared a national park. Tamang
had recently begun studying for a doctorate at Michigan State University under
wildlife biologist George Petrides and was looking for a research project and
funding to take him back to Nepal. As Seidensticker later recalled, a plan for
performing the Smithsonian's tiger study in Chitwan was hashed out between
Huxley, Tamang, and himself, as well as conservationists Raymond Dasmann
and John Milton, late one night in the bar of the Jackson Lake Lodge. When
Seidensticker revised his proposal for the study in August 1972 after further
consultation with Hornocker, he still listed India as the preferred option but
now also included Nepal as a promising alternative.[8]

That India would not ultimately be an option became clear the following
month when Ripley visited New Delhi to negotiate directly with high-level
Indian officials. Few if any of them were convinced by Ripley's argument that
the Smithsonian should not be punished for the United States' foreign policy
or for its own history of collaboration with American military and intelligence
services. When Ripley told the Indian minister of planning that the Smithsonian
was as uninvolved in the politics of Washington as the Vatican was in those of
Rome, Ripley reported to his staff after returning to Washington, the minister
"looked at me archly, and said 'so I can consider that I am now meeting the
"Pope,"'" at which we all had a good laugh." But good humor between an Indian
minister and an American pope of science did not translate into permission for
the tiger study or any other Smithsonian project in India. Ripley's conclusion
was that although all of the Indian officials had been "extremely friendly" and
cordial, they were unlikely to approve any projects except those that they had
suggested themselves, and the use of PL 480 funds was out of the question.[9]

As negotiations with the Indians stalled, plans for Nepal moved forward. In
October 1972, almost immediately after Ripley's return from India, Seidensticker and Tamang began making arrangements for a preliminary trip to Nepal.
By this time Hornocker had told the Smithsonian that he would be willing
to advise the tiger project informally but wanted his name removed from any

official documents. Seidensticker later speculated that his relationship with his mentor had deteriorated because of professional competition; once he began presenting his findings on mountain lions at professional conferences, "suddenly there were two people who knew about mountain lions." In any case, Hornocker's attention was soon focused elsewhere. In December Tamang took Seidensticker on a whirlwind tour of government offices in Kathmandu. It was much like the trip Seidensticker and Huxley had taken to New Delhi the previous March, with the crucial difference that within a few weeks they were in possession of an official letter of approval for a tiger study in Chitwan. One of their key supporters within the Nepalese government was Hemanta R. Mishra, a young forestry officer who had recently collaborated with Scottish wildlife biologist Graeme Coughley to census Nepal's population of greater one-horned rhinos. Though that collaboration had not been without its tensions, Mishra remained enthusiastic about the prospect of foreign partnerships and helped smooth Tamang and Seidensticker's path through the Nepalese bureaucracy despite some of his colleagues' resentment toward Tamang for having abandoned the forestry service.[10]

Smithsonian administrators were far less happy with the agreement between the Nepalese government and the "Smith Sonian Institution" than Tamang and Seidensticker had expected, demanding extensive revisions before they would release funds for the project. These revisions would eventually require nearly nine months of tortuous negotiations and a personal visit to Kathmandu by Huxley's assistant Ross Simons. Meanwhile the Smithsonian's enemies in Nepal and in the international conservation community were mobilizing against it. IUCN and WWF International opposed the Smithsonian's work in Nepal for much the same reasons they had opposed it in India—namely, that it was a waste of money more properly devoted to urgent conservation needs and, to the extent it would depend on funds from Operation Tiger, an act of defiance on the part of WWF-US. Their position was strengthened by support from John Blower, the parks and wildlife adviser in Nepal for the United Nations Food and Agriculture Organization, who had helped the Nepalese government establish Royal Chitwan National Park. As Mishra later recalled, Blower and his successor, Frank Poppleton, were among many "East African colonial types" who had migrated to the United Nations and other international organizations such as IUCN in the 1960s. Scientific research was peripheral at best in their vision of conservation. Seidensticker similarly recalled that the UN conservation advisers in Nepal were "ex-military officers" from British East Africa, whose "idea of conservation was barbed wire and bullets—you know, you put up the fence, you shoot the bad guys, control the perimeter, and don't let the scientists get in your [way]."[11]

In January 1973, after meeting with Tamang to discuss the proposed study, Blower began trying to convince the Smithsonian and the Nepalese government that it should not be conducted in Chitwan, if it was to be conducted at all. Blower warned Huxley that even though Tamang and Seidensticker would make an "ideal team" and "between them will no doubt produce a really first rate study," Chitwan was hardly the best location for research. The park was "subject to very considerable disturbance by local villagers and cattle, and also by tourists," and it contained probably no more than twelve to fifteen tigers. The Karnali Reserve in Bardia, a more remote site in the west of Nepal, contained a larger tiger population and was "much less disturbed." Moreover—echoing arguments that had been used against the study in India—Blower suggested that the intensive ecological study of a single population proposed by the Smithsonian was less urgently needed than a quick survey of the conservation needs of tigers throughout Nepal. Blower himself had, he told Huxley, recently proposed just such a survey to WWF International.[12]

Though Ripley recognized that Smithsonian-sponsored tiger research was impossible for the moment in India, he continued to hope that a successful pilot project in Nepal would convince the Indian government to allow the Smithsonian into its tiger reserves. In the meantime, he urged WWF-US to fund the Nepal project despite the disapproval of WWF International. Though increasingly successful at raising funds for wildlife conservation, WWF-US was mired in administrative confusion, as was WWF International; Ripley would later complain to Huxley about the organization's "general amateurish quality." In the spring of 1973, Ripley succeeded in forcing WWF-US's unpaid president, C. R. Gutermuth, to resign after Gutermuth was elected to the presidency of the National Rifle Association. On April 10, Gutermuth told the WWF-US board of directors that he was reluctantly stepping down because "some people mistakenly view the aims of the WWF and the NRA to be conflicting." Less than a week later, fully aware that by doing so it was risking WWF International's wrath, the WWF-US board approved a grant of $30,000 to the Smithsonian tiger project. The project would, according to the proposal approved by the board, "allow the U.S. to fulfill its commitment to protection of the Nepal tiger and would also provide scientific data where none exists."[13]

A few days before he learned that the grant had been awarded, Seidensticker told Huxley that he had been rethinking the way the proposal was framed. He had modeled it closely on Hornocker's work on mountain lions in Idaho and the Craigheads' work on grizzlies in Yellowstone, but his subsequent conversations and negotiations with Indian and Nepalese government officials, with the "endangered species 'heavies'" at IUCN and WWF International, and with wildlife managers in the United States had convinced him that "something is lacking. There is some barrier that we are not crossing. We are not turning people on." Instead of focusing on "'natural populations'—whatever those are," Seidensticker was now convinced that wildlife biologists needed to focus on the biological, technical, and socioeconomic constraints that kept populations of endangered species from flourishing. It was obvious that "land use in south Asia will not be based on aesthetics but on immediate economic and social need. If endangered species are to warrant any consideration in land use planning, the socio-economics cannot be ignored." Though he believed the proposal he had written was scientifically strong, future proposals would need to consider technological and socioeconomic constraints on endangered species along with the biological constraints. Doing so, Seidensticker wrote, would highlight the urgency of field research, shift proposals from the passive to the active voice, and carry "the ball one step more for the management people" who often questioned the value of basic research. This was essential because time was running out for tigers and other endangered species in South Asia: "We have got to get with it, or there won't be anything to get with."[14]

Despite recognizing the importance of a broad, socioeconomically informed approach to wildlife conservation, Seidensticker continued to frame the Nepal tiger study to prospective funders much as he had in his original proposal— as an intensive biological study of a single celebrity species. In August 1973, he applied to Leonard Carmichael, head of the National Geographic Society, for supplementary funding for the project. Perhaps calculating that exoticism would appeal to National Geographic, Seidensticker quoted E. A. Smythies, Nepal's former conservator of forests, who had described Chitwan in 1942 as "the acme of big game shooting, reserved for the sport of the Maharaja, and his distinguished guests, an Emperor, a Prince, a Viceroy." There were far fewer tigers in Nepal than in India, Seidensticker admitted, but the "cooperative and hospitable atmosphere" the Smithsonian had encountered there was just as important as tigers to the success of the project. The Nepalese government's enthusiasm for the project had been demonstrated by its approval of radio tracking, "a sensitive issue everywhere on the Indian subcontinent." As in earlier proposals, Seidensticker emphasized that immobilizing drugs and radiotelemetry would be "key to obtaining the requisite data from the tiger, as it was from the grizzly and mountain lion." Compared to other techniques, Seidensticker later recalled, radio tracking "gave you access to an animal all the time versus when the animal wanted you to have access, or just incidentally gave you systematic observations, and that opened up a whole different world of animal ecology."[15]

As the Smithsonian and the Nepalese government finalized their agreement for tiger research in early September 1973, they came under attack from Nepal's growing wildlife tourism industry. The Tiger Tops Jungle Lodge, the only tourist operation in Royal Chitwan National Park, had been founded in the mid-1960s by a Texan named John Coapman, who had visited on safari what was then still a hunting reserve and had been inspired to open a hotel there on the model of Kenya's famous TreeTops lodge. From the beginning, Tiger Tops sought to create an atmosphere reminiscent of colonial-era big game hunts, although the hunting was now done with cameras rather than guns. During the day, guests were led through Chitwan on elephant-back to view rhinoceroses and other wildlife; at night they were invited to watch from behind blinds while wild tigers fed on water buffalo that had been staked out as live bait near the lodge.[16]

After falling into debt in 1971, Coapman had been forced to relinquish Tiger Tops to a group of English partners that included Jim Edwards, a former Pan Am executive. Edwards and the lodge's research director, Charles McDougal, were deeply concerned about the hands-on methods proposed by Tamang and Seidensticker. On September 8, 1973, before the final agreement between the Smithsonian and the Nepalese government was signed, McDougal told the head of the Nepalese forestry department that he and Edwards generally supported research on the park's tigers but that they had "certain reservations owing to the fact that some of the methods which the [Smithsonian] team plans to use will necessarily disturb the tiger population to an extent which will jeopardize the role of the tiger as a tourist attraction in Chitwan. . . . From the standpoint of Tiger Tops Jungle Lodge, the results could be disastrous." Tigers that had been hunted, trapped, drugged, and collared might become skittish around humans and harder to attract to baits, McDougal suggested; moreover, even if they did continue to feed on the lodge's buffalo, "a tiger with a radio hanging around its neck is no longer an attraction." If the government did decide to approve the study, McDougal asked that it restrict the researchers from collaring the lodge's "resident tigers," a male and two females with cubs who regularly appeared at the Tiger Tops bait stations. Blower, the outgoing UN Food and Agriculture

Organization adviser, and his successor, Poppleton, seconded McDougal's concerns. A few days after McDougal sent his letter, Blower told Mishra that he believed "that the fears expressed by Tiger Tops concerning the darting of tiger with immobilizing drugs in the Chitwan Park are very valid." As he had before, Blower suggested that the project should instead be carried out in the Karnali Reserve in isolated Bardia, where it would not clash with the needs of the tourism industry. If it had to be in conducted Chitwan, tigers should be tagged only with permission of the warden and only in remote areas of the park, far from tourists and Tiger Tops.[17]

On September 19, 1973, despite the complaints of Tiger Tops and the UN advisers—which were accompanied by reminders about how much money Tiger Tops brought directly and indirectly to Nepal—the Nepalese government sent a signed copy of the final agreement to the Smithsonian. The agreement gave the Smithsonian project permission to operate in Chitwan for three years, stipulating that Tamang and Seidensticker would "help to estimate tiger population in the Kingdom of Nepal" and that the Nepalese government would be informed in advance of decisions to capture or release animals. In late November 1973, with Seidensticker and Tamang already preparing to collar their first tiger in Chitwan, WWF International made a desperate attempt to block the project with an "urgent" cable to WWF-US demanding that it suspend payment to the project, which it argued was draining money away from higher-priority conservation needs and could have serious political repercussions. It also cabled the Nepalese government with a threat to withdraw its offer of $132,000 for tiger conservation work in Nepal that had been raised through Operation Tiger if the Smithsonian project was allowed to go forward. Pretending to have misread the cable and acting without the knowledge of his superiors in Nepal's Parks and Wildlife Conservation Office, Mishra told WWF International that the Smithsonian project had been approved and that Nepal would happily accept the proffered Operation Tiger funds.[18]

Mishra later recalled that the Nepalese government welcomed the Smithsonian project both because the Nepalese were sympathetic to the United States and because they saw competition between various international aid organizations as an opportunity: the more organizations competing to work in Nepal, the more money, equipment, training, and advice Nepal could expect to garner. Though staking his career on parks and wildlife rather than the well-established field of forestry was a risky move, Mishra believed the project would benefit his country and his own interests. Partly because of family connections, Mishra was able to convince the Nepalese royal family of the project's value, particularly Prince Gyanendra Bir Bikram Shah, with whom he had collaborated to establish Royal Chitwan National Park and whom he would later describe as "a staunch traditionalist and the force behind Nepal's nature-conservation program." For Seidensticker, writing to Huxley from the field camp in Chitwan in early December, the Nepalese support for tiger research provided an opportunity that had been tragically lost in India. The opposition from WWF International and IUCN "together with India's backward approach shows that as far as Project Tiger goes something is really lacking. That something is the basic ecological information on the tiger that our study is designed to provide."[19]

The Vital Sense of Partnership and Full Collaboration

Mishra later recalled that when Seidensticker arrived in Nepal to begin fieldwork with Tamang, he carried himself like a "Montana cowboy," and for good reason. Seidensticker had grown up on a large cattle ranch in southwestern Montana, learning to hunt with his father, a physician in whose professional footsteps he considered following well into his university years. Seidensticker recalled that it was the Craigheads' article "Knocking Out Grizzly Bears for Their Own Good," which appeared in *National Geographic* when Seidensticker was still in high school, that first suggested the alternative career he would eventually pursue. As an undergraduate at the University of Montana, Seidensticker was hired by John Craighead as a field assistant for the grizzly bear project. After graduating in 1966, he continued to work with Craighead, first studying the effect of DDT on raptors for a master's degree and then spending six months in Yellowstone radio-tracking elk for the "big game" study funded by the Atomic Energy Commission. It was during this latter period that Seidensticker read George Schaller's pathbreaking 1967 book, *The Deer and the Tiger*, which was based on research in Kanha National Park in central India. Schaller's innovative field methods were an inspiration for Seidensticker, as they were for many other wildlife biologists of his generation; it was only after reading the book that he started "really being a biologist," he later recalled. Meanwhile, Seidensticker was getting a firsthand introduction to the complex politics of wildlife research through long conversations in car rides to and from Yellowstone with John Craighead, who was then battling with the Park Service over how grizzly bears should be managed and studied.[20]

Many of the research techniques that Seidensticker learned in Yellowstone and in his subsequent doctoral research with Hornocker on mountain lions in Idaho could be applied with little change in Nepal. The Craigheads had refused to allow their radio-tracking equipment or designs to be used for the mountain lion study, so Seidensticker had been forced to build his own collars and tracking gear with the help of a radio engineer in Moscow, Idaho, named A. R. Johnson. It was this equipment that eventually made it through Nepalese customs in the fall of 1973 for use on tigers. The compressed-air gun used for darting was also the same, although the specific immobilizing drug differed. Instead of succinylcholine chloride, the muscle relaxant the Craigheads had initially used on grizzlies in Yellowstone, or phencyclidine hydrochloride, which many biologists began using in the late 1960s because it appeared to have a wider margin of safety, Seidensticker and Tamang used a compound called CI-744 that was produced for experimental purposes by the pharmaceutical company Parke-Davis. CI-744 was a mixture of a dissociative anesthetic and a tranquilizer that took effect within minutes, had relatively minor side effects, and allowed quick and predictable recovery, all of which were important considerations when working with highly valued and endangered animals in difficult terrain.[21]

While darting and collaring equipment remained largely unchanged, locating and immobilizing tigers in the thick vegetation of Chitwan demanded an entirely new set of techniques. In Idaho, Seidensticker had adopted a method of capturing mountain lions that Hornocker had developed in partnership with a professional cougar tracker named Wilbur Wiles, who used a team of trained dogs to force the cats into trees, where they could easily be darted. Tigers were much larger and more dangerous than mountain lions, however, and the tall, dense grasses of Chitwan made any approach on foot perilous. Several years into the project, Tamang would be pulled out of a tree and seriously mauled by one tiger. As in Idaho, local trackers, trappers, and hunters provided expertise that could be adapted to the needs of wildlife research. In Idaho the requisite expertise had been nurtured by a tradition of predator control by and for ranchers; in Nepal it had been nurtured by a tradition of colonial and aristocratic hunts. The head shikari—hunter or hunting guide—for most of the duration of the Smithsonian project was Prem Bahadur Rai, whom Mishra later recalled as the best tiger tracker in Nepal. Prem had spent nearly two decades working as a shikari for the Nepalese royal family before taking a job with Tiger Tops, where he worked for several years before accepting a position with the Smithsonian project. Prem's expertise in reading tiger tracks and signs, choosing the best sites for staking out live buffalo calves as bait, and coordinating the elephant drives that were used to corner tigers so they could be darted—a task aided by the use of long white strips of cloth to funnel the tigers toward the shooter, who was usually hiding in a tree blind—was crucial to the success of the project.[22]

Even with expert trackers and well-trained elephants, collaring tigers was still dangerous work. At the beginning of December, in one of Seidensticker and Tamang's first attempts, the female tiger they had surrounded charged an elephant named Rabi Kali, who bolted for two hundred yards at full speed while Seidensticker and the elephant driver did their best to hang on. After a month, Seidensticker and Tamang had managed to collar a leopard and a tiger, with heavy reliance on the collective tiger-hunting expertise of the team of shikaris and what Seidensticker described as "the best of the best government elephants," which had been temporarily loaned to the project. As Seidensticker explained to Huxley, "Catching tigers on a regular basis is not a one man sneaking around in the grass with a dart gun operation," but fortunately Tamang was "a tiger hunter of the first order and that is the kind of thing we need to know to catch tigers." Seidensticker also told Huxley that his idol Schaller, who had recently had a frustrating experience attempting to radio-track lions in the Serengeti and was now hoping to have more success with snow leopards in Pakistan, had told him he was impressed at the Chitwan project's progress. The key to their success was the blending of Seidensticker's radio-tracking expertise with the tiger-tracking expertise of Tamang and the shikaris. When Seidensticker and Tamang wrote a popular account of the tiger study for the Time-Life Science Yearbook the following year, they noted that "detailed ecological information on this great cat can only be gathered by scientists using the most modern field techniques," such as immobilizing drugs and radiotelemetry, but that "the key to mobility in the thickly vegetated areas where tigers live, is the use of well-trained elephants—one of man's oldest domestic animals."[23]

Although the collaboration between Tamang and Seidensticker was vital to the success of the project, it deteriorated rapidly under the often stressful conditions of cross-cultural fieldwork. As early as October 1973, before fieldwork had even begun, Seidensticker complained to the Smithsonian about Tamang's demands for a pay raise, which he thought were excessive for a doctoral student, even one with a wife and children to support. Tamang, meanwhile, resented Seidensticker's attempts to take the leading position in the collaboration despite Tamang's superior expertise with regard to Nepal, Chitwan, and tigers. Throughout the winter and spring, the two struggled for control over the project, while personal conflicts among the project staff and their families grew and the efforts

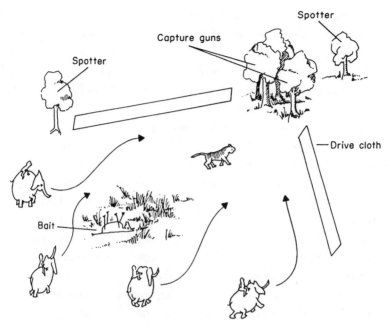

The technique used by the Smithsonian-Nepal Tiger Ecology Project to dart tigers was similar to techniques that had been used in earlier tiger hunts. After a tiger had killed the bait—usually a young water buffalo—elephants drove the animal into a funnel created by two long stretches of white cloth, at the end of which a shooter was waiting with a dart gun. (Source: Melvin E. Sunquist, *The Social Organization of Tigers* [Panthera tigris] *in Royal Chitawan National Park, Nepal* [Washington, DC: Smithsonian Institution Press, 1981], 15, figure 11, courtesy of the Smithsonian Institution Scholarly Press)

of Tiger Tops and the international conservation community to undermine the project continued. Tamang eventually retreated to Kathmandu, while Seidensticker continued trying to collar and track tigers at Chitwan.[24]

By the summer of 1974 the Smithsonian had heard enough complaints from both parties to decide that the partnership was a failure and that Seidensticker, the more easily replaced of the two, would have to go. In July, in a letter copied to Mishra and Tamang as well as to Smithsonian staff and the WWF-US board, Huxley informed Seidensticker that he was being immediately removed from the project, though the Smithsonian would continue to support him for a few months while he wrote up his findings. The value of Seidensticker's scientific contributions was not in question, Huxley explained; the problem was his apparent reluctance "to genuinely resolve what has been a very serious problem of not developing the vital sense of partnership and full collaboration with the Nepalese involved in the project." A second factor in the decision was that the Smithsonian intended to shift the focus of the project from a short-term study of tigers to the establishment of a long-term research presence in Nepal. For that purpose, Huxley explained, it would be preferable to have an American collaborator who was more familiar with the Smithsonian's guiding principles and institutional organization than Seidensticker was. Tamang was to take over as sole principal investigator by the end of the month, and a new American co-principal investigator would join the project in the fall.[25]

The Smithsonian's interest in establishing a long-term presence in Nepal was tied to its growing conviction that its relations with India were beyond repair. Over the summer of 1974, there seemed to be an opening for improved relations between the Smithsonian-Nepal tiger project and India's Project Tiger when Sankhala asked the Nepalese government for permission to visit the tiger project in Chitwan. The Nepalese government agreed only on the condition that the Smithsonian researchers be allowed into India, however, and when Sankhala refused, it withdrew the invitation. Smithsonian researchers subsequently refused even to send copies of their progress reports to Sankhala. Another opening came late that summer, when Seidensticker accepted an invitation to help immobilize and translocate a man-eating tiger in the Sunderbans, the extensive mangrove forest stretching across coastal areas of West Bengal and Bangladesh. Although it took several days for Seidensticker to dart the animal, the translocation appeared to be going smoothly until the tiger was found dead almost immediately after being released. Sankhala blamed Seidensticker for the death, telling the press that the tiger had been so weakened by the darting that it had been killed by wild pigs, while Seidensticker argued that another male tiger, most likely defending its territory, was to blame. The Indian press largely took Sankhala's side, reporting, Seidensticker later recalled, that he had "slinked back to Kathmandu" in defeat. In any case, any possibility of what the historian Lewis describes as a "tiger détente" between the Smithsonian and India had been lost.[26]

Seidensticker's replacement, Melvin E. Sunquist, had also hoped to study tigers in India but was forced by the Smithsonian's poor relationship with the Indian government to work in Nepal. In early September 1974, in advance of Sunquist's arrival at Chitwan, Huxley's assistant Simons assured Tamang that he was "highly qualified and even tempered," in implicit contrast to Seidensticker. Moreover, Sunquist was familiar with the Smithsonian's goals and operating procedures in international research, having recently spent several years radio-tracking sloths at the Smithsonian's field station at Barro Colorado Island in Panama. He was now studying for his doctorate under Siniff at the University of Minnesota, which Simons described to Tamang as "a leading center" for radiotelemetry in the United States. With Sunquist's arrival in Nepal in October, the Smithsonian-Nepal Tiger Ecology Project began using radiotelemetry gear from the Cedar Creek Bioelectronics Laboratory rather than the equipment that Seidensticker and Johnson had developed in Idaho, but otherwise the project's methods and aims remained virtually unchanged. Mel Sunquist's wife, Fiona Sunquist, who soon joined him in Chitwan, later recalled that he had accepted the assignment with reluctance, fully aware that the previous collaboration had ended badly and "that reputations, his own as well as those of the project's major supporters at the Smithsonian, were at risk."[27]

Before Sunquist's arrival the project had been relying exclusively on WWF-US funds, which limited research to subjects directly relevant to tiger conservation and provided an unreliable foundation for long-term planning. In December 1974, David Challinor, who had replaced Galler as the Smithsonian's assistant secretary for science in 1971, explained to Ripley that the Smithsonian's short-lived Environmental Programs Office had been eliminated because of the increasing politicization of environmental issues. A side benefit of the closure was that it freed up funds that could be applied to the Smithsonian's core research mission, including its work in foreign countries, which had been scaled back because of the loss of funds from the Department of Defense and PL 480. In 1975, this would make it possible for the Smithsonian to contribute $40,000 to the Nepal tiger project in addition to the $40,000 it was already receiving from WWF-US.[28]

Despite these changes, the radio tracking of tigers remained the focus of the Smithsonian-Nepal Tiger Ecology Project. Seidensticker's and Tamang's collaboration in Chitwan had been too short and too compromised by personal and professional conflict for them to reach to any significant conclusions about the tiger population or its prey. After Seidensticker's departure Tamang continued to focus on prey species as the subject of his doctoral dissertation for Petrides at Michigan State, while Sunquist expanded and elaborated upon the investigations that Seidensticker had begun into the tigers' "land tenure system." Territoriality had been a central concern in Hornocker's work on mountain lions, as it had been in Schaller's research on tigers in India. Hornocker had concluded that mountain lions had "a fairly rigid system of territorialism," a finding that Seidensticker's radiotelemetry study had largely confirmed. Schaller had concluded that the tiger's land tenure system resembled that of the African lion—that is, each female defended an area containing the prey she needed to survive and feed her cubs, while each male defended an area containing a number of females with whom he might mate. Still, Schaller's conclusions had remained tentative, since he had never witnessed actual "territorial behavior," such as fighting at the border between two territories, among the tigresses at Kanha. Seidensticker had also been influenced by the work of John F. Eisenberg of the National Zoological Park, who had argued that territorial behavior could not be understood apart from its evolutionary and ecological context. The existence of an evolved "territorial imperative" was not enough to predict a predator's land tenure system, Eisenberg had argued; one also needed to account for the availability of prey and other environmental factors.[29]

Sunquist drew on all of these sources and on the work of his adviser Siniff, who had a long-standing interest in statistical models of animals' use of space. Based on his research at Chitwan, Sunquist would eventually conclude that the tigers had a territorial system based on "prior rights"; as a rule, once a tiger had established control over an area, no other tiger would be able to establish permanent residence there until the original tiger had died or moved away. Moreover, tigers' ranges appeared to be true "territories" in the sense that they were exclusive and aggressively defended against other tigers of the same sex. The territories were often so large, however—especially in the case of the males—that it might be weeks before a tiger revisited a particular portion of its range. Tigers kept their rivals out of these extensive territories through marks of ownership, including scent marks, feces, and scrapes, and through threats of attack. The more successful a tiger was at finding food, the more energy he or she could devote to marking and defending territory, including making what Sunquist called "exploratory probes" into neighboring tigers' territories when they had been left unmarked for an unusual length of time. While tigers' territorial behavioral was shaped by their evolutionary history, Sunquist noted that human alterations of the landscape had a significant impact on the shape and size of tiger territories. The particularly large territory of one of Chitwan's male tigers was probably the result, he surmised, of park roads and trails that made it easy for him to cover large distances, bait stations that reduced his need to spend time and energy hunting, and borders on agricultural lands that did not need to be defended from other tigers.[30]

Tiger Tops was just as territorial as any tiger, and its territorial markings far easier for the Smithsonian researchers to read. Soon after the project began, the researchers came to an informal agreement with Edwards and McDougal not to radio-collar any tigers west of the Sukhibar guard post, located about half-way between the Smithsonian camp at the east end of the park and Tiger Tops at the west end. The agreement forged at the "Sukhibar Summit," as Mishra later called the meeting where the dividing line was established, depended on a shared assumption about tiger territoriality—namely, that none of the lodge's "resident" tigers, those who appeared regularly at bait stations and were most likely to be seen and photographed by tourists, ranged east of the guard post. That assumption was proven wrong in February 1976, when Sunquist collared a large male tiger just east of Sukhibar with the help of James L. David Smith, a graduate student at the University of Minnesota who was considering joining the project for his doctoral research. The tiger, already known to the project as 102, had been darted and collared by Tamang several years earlier but never successfully radio-tracked. After the collaring Smith and Mel and Fiona Sunquist stopped by Tiger Tops to pick up a bottle of rum for that evening's celebration of the successful collaring, which was when, Smith told Simons in a letter from Nepal soon after, the "melodrama" began. When they showed McDougal a Polaroid photograph of 102, he immediately recognized him as the "Dakre Tiger," the largest of the lodge's resident males.[31]

As a retrospective account by Mishra and two of the Smithsonian administrators of the tiger project put it, while the discovery that the Dakre Tiger, 102, ranged all the way from Tiger Tops eastward past the Sukhibar guard post "was an illuminating discovery biologically, it was an inevitable faux pas of diplomacy." It was exactly the result that the agreement at the "summit" had been intended to avoid, and which Tiger Tops management had feared since first learning about the project. Smith told Simons that while he thought McDougal was "sympathetic" to the Smithsonian project and willing to forgive the collaring of 102 for the sake of science, his opinion carried little weight with the Tiger Tops management, who, in Smith's opinion, didn't "give a shit about the future of tigers—their only concern is showing the tourist what *they* think the tourist wants to see." He expected they would "make quite a fuss" as soon as 102 or any other collared tiger appeared at the bait stations.[32]

However sympathetic McDougal may have seemed in private conversations with Smith and other tiger study staff, he was a loyal advocate of his employer's cause with Mishra and others in the Nepalese government. In early March 1976, in response to McDougal's complaints about the collaring of 102, Mishra demanded that he stop trying "to drag me into" the conflict. The Nepalese government had never, he told McDougal, agreed to limit the Smithsonian project's reach within the park, and "whether you show tigers to the tourists collared or uncollared or whether you put a stop on showing tigers is hardly my business." Mishra also disputed McDougal's claim that the tiger project was making it impossible for a film crew sent by the British company Survival Anglia to finish a film on tigers in the park. In his unwanted role as a mediator between Tiger Tops and the Smithsonian, Mishra was far from a disinterested party. Not only had he helped shepherd Tamang and Seidensticker's initial proposal and the subsequent Smithsonian renegotiations through the complexities of Kathmandu bureaucracy and palace politics, but a month earlier, with Tamang nearing the completion of his doctorate at Michigan State, Simons and Smith had written to Mishra to tell him that they hoped he would take Tamang's place as the tiger project's principal investigator.[33]

From Ripley's perspective, Tiger Tops' opposition to radio collaring was based on the same misguided philosophy of wildlife conservation that had kept the Smithsonian study from taking place in India—namely, a kind of mystical belief that endangered wildlife could be preserved only by completely isolating it from any sort of human influence. A few days after Mishra rejected McDougal's request to intervene in Chitwan, Ripley sent a report on the project's recent woes to Challinor, Tamang, and Godfrey Rockefeller, one of WWF-US's board members. Tiger Tops and its supporters in the international conservation community, he explained, "apparently resent a scientific technique brought in by Americans, which through the presence of 'collars' on tiger or leopard, might delude the tourist into feeling that the wild animals are not wild, but rather escaped pets or circus strays." That point of view was, in Ripley's opinion, "parochial and rather demeaning to the intelligence of the tourists." While Tiger Tops' attacks were unlikely to succeed, given the level of support for the project in the Nepalese government, they were a constant distraction. Such opposition to research in Nepal was particularly dispiriting because India's Project Tiger, despite its success in creating strictly protected tiger reserves, had generated "not a trace of hard data" about the status of the tiger populations within them. Indian officials such as Sankhala had succeeded in doing on a national scale what Tiger Tops hoped to do in Chitwan: "Other nations' programs in sophisticated wildlife conservation are often impeded by such purists' beliefs that endangered wildlife can only be preserved if it is untouched by human hands, forgetting conveniently that the same species have been brought to their present plight by those same human hands. Like believers in spiritual healing they prefer to divorce themselves from the sciences, medical [*sic*], or in this case applied ecology and zooculture."[34]

However misguided the opponents of radio tracking might be, Ripley admitted that the tiger project in Chitwan had not yet realized the ambitious goals with which it had been initiated. By the spring of 1976, the project had radio-collared seven tigers, four leopards, two sloth bears, and several sambar deer, but in Ripley's assessment it had not yet achieved its primary goal: "to develop methods for determining the number of tigers" within the park and, eventually, throughout Nepal and India. Still, in Ripley's opinion radio tracking remained far more promising than the "pugmark" system touted by Indian forestry officials, in which paw prints and other kinds of tiger signs were used to identify individual animals. McDougal's attempts to identify individuals using photographs of facial stripe patterns were somewhat more promising than pugmark tracking, Ripley thought, but the technique seemed limited to tigers who could be attracted to baits, since it was only there that the high-quality photographs necessary for accurate identification could be taken. In any case, Ripley concluded, despite its limitations as a means of censusing tigers, the radiotelemetry project was already providing significant insights into the tigers' "day-to-day life" that might aid in conservation efforts.[35]

The leadership of Project Tiger was well aware of Ripley's belief that that India's methods of counting its tigers were flawed. A week before Ripley sent his attack on wildlife "purists" to Challinor, Tamang, and Rockefeller, Sankhala told the Indian press that the cheap, efficient, and indigenous "pugmark tracer" system was far superior to the American "radio collar" system. The latter was expensive, risky, and particularly unsuited for studying tigers, Sankhala claimed, and it was only Ripley's ignorance or prejudice against India that prevented him from recognizing its advantages. As the author of a sympathetic article in the *Times of India* paraphrased Sankhala's position, "The philosophy behind Project Tiger in India was not to catch or collar tigers but to allow them an undisturbed habitat to enable them to grow in numbers."[36]

Still, because much of its funding came from foreign donors and from organizations largely staffed by scientists, such as WWF and IUCN, Project Tiger could not entirely exclude foreign experts or ignore increasingly common techniques such as radio tracking. In the spring of 1976, it invited two foreign scientists affiliated with IUCN to conduct a review of India's tiger conservation efforts and asked David Mech and another wildlife expert from Minnesota— Ulysses Seal, a specialist on captive breeding of endangered species at the Minnesota Zoo—to lead a workshop in northern India on tranquilizing and telemetry methods. After a basic introduction to the techniques at the forestry school in Dehradun, Mech and Seal conducted field demonstrations at several wildlife sanctuaries. The radiotelemetry demonstrations led by Mech were conducted at the Ranthambore Tiger Reserve in Rajasthan, and the participants in the course included the managers of a number of other tiger reserves, including Corbett, where the Smithsonian had hoped to carry out its tiger study, and Kanha, where Schaller had conducted his research in the mid-1960s, but no tigers were handled. Instead Seal and Mech confined their demonstrations to deer and other prey species that were far easier to handle, both logistically and politically, than the controversial tiger.[37]

If the invitation to Mech and Seal indicated a new openness to hands-on techniques among the leadership of Project Tiger and the Indian Forestry Service, it was short lived. Once the workshop on tranquilizing and telemetry was over, reserve managers continued to rely on the pugmark system as they had before, and no one attempted to put into practice what the Project Tiger report on the workshop had described as "modern disciplines in wildlife management." Sankhala remained committed to a strict protectionist approach to tiger conservation in which the kinds of research techniques used on a daily basis by the Smithsonian-Nepal Tiger Ecology Project—whose scientists were conspicuously not invited to the workshop, even though they had more experience immobilizing and radio-tracking tigers and their prey than anyone—had no place. At a WWF-sponsored conference in the fall of 1976, Sankhala confidently declared that "the Tiger has been Saved"; when he stepped down as head of Project Tiger the following year, he noted that India had accomplished this remarkable feat while also maintaining the tiger's "dignity and not allowing him to be degraded to the status of a trophy, a guinea pig or a frog in a biology class."[38]

In the aftermath of the collaring of the Dakre Tiger, an uneasy truce between Tiger Tops and the Smithsonian project was gradually reestablished. When Siniff visited Chitwan a little more than a month after the collaring, he found the Tiger Tops management still "very upset" about the collaring of the Dakre Tiger and the prospect of additional "residents" being collared. As far as Siniff could make out, he told Simons after returning to the United States, their concerns "centered on the aesthetics of the tourists viewing a tiger with a collar," even though they admitted that less than 10 percent of visitors were likely to be offended. That number would still, they argued, have a significant effect on their bottom line. Siniff tried to convince them that a good interpretive program—including, perhaps, a display of radiotelemetry gear at the lodge—would make the project "less distasteful" to their clients, but, he told Simons, "they did not jump at the chance." For the moment, the issue was moot, since the Smithsonian project had no intention of collaring additional tigers near the lodge or of removing 102's collar. Moreover, given the continued staunch support for the Smithsonian project from Mishra and from Prince Gyanendra, Tiger Tops' complaints had little force. When Smithsonian administrator Christen Wemmer visited Nepal a year later, he found that Tiger Tops' tactless assertion of its "territorial rights" in Chitwan was actually undermining its relationship with the Nepalese government; Tamang told him he believed that "if given enough rope they will surely hang themselves."[39]

In the meantime, however, the Smithsonian project had run into its own difficulties with the Nepalese government. By the spring of 1976, CI-744 had been used to immobilize tigers and leopards on dozens of occasions without incident. Given that the dose of the drug had to be calibrated to each animal's weight and condition, which were often estimated by eye under less than ideal conditions, the record was a sign of the drug's relative safety. The first fatal overdose occurred in late April 1976 about six miles west of the project's main camp, which was located near the village of Sauraha at the east end of the park. After apparently starting to recover from the drug, the 255-pound female tiger that had been darted died suddenly, leaving behind three approximately six-month-old cubs, each weighing about a hundred pounds.[40]

As soon as it learned of the incident, the Nepalese government placed an immediate moratorium on further darting and suppressed any reports about the death in the Nepalese press. After a week of debates within the government over what to do with the cubs, Sunquist sent a letter describing the incident to Simons "in strictest confidence." The cubs had been given buffalo carcasses to feed on while their fate was decided, and Tamang had proposed continuing to feed them until they dispersed on their own. Sunquist, however, believed the cubs were too young to survive without their mother; after dispersing they would probably either die of starvation or be killed by adult tigers. In the dense vegetation of Chitwan it would be nearly impossible to know whether they had survived, which was why, Sunquist suspected, Tamang's proposal appealed to government officials; "it in effect removes the responsibility." A day after Sunquist posted his letter to Simons, Tamang wrote to Ripley to let him know that the palace committee had decided to follow Sunquist's advice. A twenty-one-acre enclosure was built near the temporary camp where the cubs' mother had been darted using the same white cloth used to direct tigers in darting operations, and the cubs were driven inside it. The cubs and the buffalo carcasses they were feeding on attracted mature tigers, however, particularly a large female known to the project as 106, whose constant roaring eventually drove the frightened cubs to flee the enclosure, after which it was impossible to recapture them. By the beginning of June at least two of the cubs were still alive, but Simons was "skeptical of the final outcome."[41]

In the summer of 1976, as the term of the three-year agreement that had been signed by the Smithsonian and the Nepalese government in September 1973 neared its conclusion, the two parties began negotiating a new agreement. The Smithsonian's goals in the negotiations were to broaden the scope of research in Chitwan and to lay the groundwork for a permanent research station. In August 1976, Simons told Wemmer, who had just been appointed to the project, that the Smithsonian had been trying to expand its work in Chitwan beyond tigers since 1975, when it began contributing funds to complement those provided by WWF for tiger research, but that the Nepalese government had proved reluctant to approve or support research projects that were not immediately relevant to tiger protection.[42]

This had become particularly evident in the case of Rebecca Gay Troth, a doctoral student in botany who had come to Chitwan on a grant from WWF-US earlier in 1976 to study *Bombax ceiba*, the silk cotton tree. Although the trees played little direct role in the lives of tigers, Troth and the Smithsonian had convinced WWF-US that "plant communities are the critical foundation for the whole wild life of the area." As Simons explained to Wemmer, the Nepalese government seemed to doubt the value of Troth's work, throwing up numerous bureaucratic and logistical hurdles in the way of her fieldwork that other Smithsonian researchers had not faced. Simons believed that Troth's gender was one factor; he told Wemmer that she was "single with a clear determination to show the world that she's as good as any male botanist. That healthy attitude doesn't mesh well in a male chauvinist pig country like Nepal." The second reason and more troubling factor, from Simons's perspective, was that the relevance of her research to the conservation of the tiger or Chitwan's other high-profile endangered species, the rhinoceros, was not immediately obvious.[43]

In September 1976, the Nepalese government granted the Smithsonian a two-year extension of the tiger project on the understanding that Troth's research would not be continued. Simons continued to hope that Troth's study would be able to proceed if it could be sold "under the tiger rubric" to the Nepalese government. A few weeks after the signing of the agreement, he urged Troth, Tamang, and Smith, who was preparing to begin his fieldwork in Chitwan, to maintain a broadly ecological approach to their work despite pressure from the Nepalese government and WWF to focus on tigers. Troth's troubles and recently renewed efforts by WWF International to force WWF-US to terminate its support for Smithsonian research in Chitwan had left Smith and his adviser at the University of Minnesota, wildlife biologist Peter Jordan, deeply concerned. Jordan was familiar with the complexities of international research, having previously advised a student working on a Smithsonian project on Asiatic lions in India's Gir Forest, but the problems in Nepal struck him as unusually bad. They were, he told Challinor in December, evidence of "either a highly unstable relationship between the Institution and the host county or some gross misunderstanding in the working agreement between the two." In either case, it was an extremely risky environment for doctoral research. Just a month before Smith was scheduled to begin fieldwork in Chitwan, they were seriously considering refocusing Smith's thesis work on antelope in Sudan, where he had previously worked as a Peace Corps volunteer. Smithsonian administrators managed to assuage Smith's and Jordan's concerns long enough to get Smith to Chitwan, but even after he began his fieldwork, frustrated by bureaucratic roadblocks that prevented him from obtaining a small aircraft for tracking tigers that had dispersed beyond the borders of Chitwan, Smith was still considering abandoning Nepal for Sudan.[44]

Throughout the spring of 1977, Smithsonian administrators sought to convince the Nepalese government to allow Troth to return to her fieldwork. In April, Challinor and Simons told Ripley that they thought the Nepalese government's decision about the matter was a test of its true commitment to a partnership with the Smithsonian. If it refused to accept her work as legitimate, they suggested, "perhaps we should consider scaling down our efforts or perhaps closing down." Ultimately, however, the Nepalese government held firm to its position, and the Smithsonian accepted that it would have to make some compromises to maintain its relationship with the host country. It was a clear victory for the Nepalese government over a powerful foreign institution carrying out research within its borders, but it also left a residue of mistrust and mutual suspicion. In June, when the hopelessness of Troth's case had become clear, Smith told Wemmer that he suspected that Tamang, a "master of the subtle remark," had doomed her project by describing her to Nepalese officials as "an American girl doing some minor project on *Bombax* so she can get her PhD." That fall the Nepalese government officially confirmed that Mishra would carry out his doctoral research at Edinburgh University as principal investigator of the tiger project, that it was interested in exploring the possibility of establishing a permanent research center at Chitwan, and that the Smithsonian-Nepal Tiger Ecology Project would be allowed to operate for at least another three years. The Smithsonian had preserved its access to Chitwan but only at the price of abandoning, for the moment, its effort to broaden the scope of research beyond tigers and their prey.[45]

Adventure-Style

When Jordan visited Chitwan late in the fall of 1977 to check on Smith's progress, he found his student distracted by a "steady flow of visitors to the project" that included Jordan himself, "wildlife biologists just passing through Nepal, citizens who have made specific contributions . . . , officials of the Government, and officers of the Smithsonian and their guests." Some of these visits were professionally useful; "others just use up time." The "worst interruption," Jordan wrote in his report for the Smithsonian, occurred "when royalty come to hunt in the Park; all research stops for several weeks, but we accept this as an unavoidable condition of working in Nepal." Although the Nepalese government had banned tiger hunting throughout the country and hunting of all animals within the national parks, Prince Gyanendra and other members of the royal family continued to hunt in Chitwan. In June, for example, Smith had spent nearly a month in Kathmandu largely, as he had told Wemmer at the time, in order "to be out of the way while the King hunted Gaur (*endangered species*) in the park." Without the Nepalese royalty's enthusiasm for big game hunting, the Smithsonian tiger study might never have gotten approved, but such interruptions reminded the researchers that their reasons for supporting wildlife conservation were not necessarily the same as those of their hosts.[46]

Research was also interrupted by visits from foreign donors. Such visits helped maintain the project's funding, but they were both time-consuming and risky, since not all visitors left with a positive impression. A month before Jordan's visit, the American ambassador to Nepal, Douglas Heck, had accompanied a group of elite WWF donors and board members on a tour of Chitwan. Francis L. Kellogg, the outgoing president of WWF-US, was one of the participants who was more dismayed than impressed by the spectacle of Smith and the team of shikaris darting and tagging one of the park's tigers. After hearing Smith explain that Chitwan contained the most intensively studied population of tigers in the world, Kellogg later told the director-general of WWF International, he

> could not help but wonder if the rare opportunity to observe and study Panthera tigris at Chitwan for Ph.D. theses had not overcome the necessity to allow this great animal such freedom from human pressure as is possible on today's overcrowded planet. To what degree, I pondered, does the radio collar affect the shy wild animal that carries it? Or the vehicles and aircraft that can so unerringly home in on its most secret lair? And the darting process, what of that? What lasting effect can shooting and drugging of a wild animal have? In retrospect, it seemed that the personal ambition of Ph.D. students may have overridden the desperate needs of an endangered species.

Heck, in contrast, told Ripley he had been thrilled to have a "chance to pat a *wild live breathing* tiger (drugged of course) and hold him by the tail. How many Ambassadors can claim that—and still be alive!" At Ripley's urging, Heck respectfully informed Kellogg that "from a strictly layman's point of view," he could see nothing wrong with the research at Chitwan, but there is no evidence that this intervention had any effect on Kellogg's opinion of modern wildlife biology.[47]

In general, however, visitors to the Smithsonian project were pleased and impressed by the experience, in no small part because of the welcome they received from Smith. Years later, Challinor would write a recommendation letter for Smith in which he praised his cheerful and competent fulfillment of one of his most important "extra-curricular" tasks in Nepal: acting as "tour guide for distinguished visitors to our Tiger Camp." Smith also helped smooth relations with the Smithsonian project's antagonists at Tiger Tops. In December 1977, Smith told Simons that he was "learning a lot" from McDougal and that Jim Edwards had offered to lend some of the lodge's elephants to the Smithsonian team for darting operations. Even Poppleton, whose opposition to the project earlier that year had been one of the reasons that Smith had seriously considered decamping to Sudan, was now helping him gain access to a United Nations airplane that could be used to search for dispersing tigers. By the spring of 1978, Heck was also crediting Smith with improving relations with the Nepalese government, as evidenced by the appearance of several positive articles in the Nepalese English-language press. "The project is doing well and the Nep-

Hemanta Mishra (facing the camera), James L. David Smith, and Prem Bahadur Rai (in the knitted cap) help weigh a tiger that has just been immobilized. (Courtesy of the Smithsonian Institution Archives)

alese attitude towards it has become far more positive," he told Ripley. In June 1978, when Mishra returned to Nepal after spending the academic year in Edinburgh, he too found that Smith's diplomatic efforts with Tiger Tops, Poppleton, and the Nepalese government had paid off and that morale among the project staff was high. In some ways, Mishra later recalled, "Minnesota Dave" was "as Nepali as any of us," even though he never learned the Nepalese language. He loved spicy food, operated on "Nepali time," and was adept at dodging the bureaucracy; on cold mornings in camp, Mishra recalled, he would sometimes join Mishra and his wife Sushma in bed with a hot beverage, a familiarity that might be taken by a relative or close friend in Nepal but was unheard-of for an American visitor.[48]

One of the most important and successful visits to the Smithsonian project was that of Jackson, the head of WWF's Operation Tiger. Jackson was sufficiently impressed to write an account of his visit for *Smithsonian* magazine titled "Scientists Hunt the Bengal Tiger—But Only in Order to Trace and Save It," which was illustrated with photographs of the darting, collaring, and radio tracking of a mature female tiger known to Tiger Tops as Chuchchi. Unlike Kellogg, Jackson expressed no reservations about the intensive study of the population or the ambitions of American Ph.D. students. Mishra thought the article was "fine" when Jackson showed him a draft during a visit to the United Kingdom; the most important thing, he told Simons, was that Jackson and the IUCN and WWF International leadership in Switzerland now seemed to have become enthusiastic supporters of the project. When Simons saw a draft, however, he told *Smithsonian* he was "troubled by its adventure-style. I expect to see natives popping out the jungle singing songs of praise to the white men." Some of "this drama," Simons suggested, "could be toned down a bit." In the published version that appeared in August 1978, Jackson compared Chuchchi's capture and collaring to an "old-style hunting story" and noted Prem Bahadur Rai's long service as a shikari for the royal family, but the "adventure-style" was relatively muted.[49]

Simons's concerns about sensationalism and colonial nostalgia extended to film and television coverage of the tiger project, though financial considerations also played a significant role in determining whether a particular film crew would be welcomed. In the spring of 1978, for example, negotiations with Marlin Perkins's *Wild Kingdom* to film the project had fallen through when Simons demanded that the producers make a financial contribution to the project on the order of the $30,000 in services it had received from Survival Anglia, whose

Prem Bahadur Rai, the chief shikari of the Smithsonian-Nepal Tiger Ecology Project, with a collared tiger. (Courtesy of the Smithsonian Institution Archives)

documentary *Tiger, Tiger* had aired the previous year. The production company's Richard Reinauer insisted that *Wild Kingdom* held a unique place in the world of wildlife film and neither could nor would pay a fee "for the privilege of filming a project." Similarly, when the producers of the American television network ABC's *American Sportsman* series approached the Smithsonian about filming in Chitwan later that year, Simons told Mishra that he had decided not to grant permission in part because of his doubts about "the appropriateness of this work appearing on a rather pro-hunting show." By April 1980, however, Simons had reconsidered. *American Sportsman* had recently increased its coverage of conservation and science, he told Ripley, including an episode on Jane Goodall's work with chimpanzees. Moreover, unlike Reinauer, ABC had agreed to "donate" $10,000 to the project.[50]

The tiger project staff in the field were well aware of the financial motivations for allowing film crews to interrupt their research. Mishra later recalled that when the ABC crew arrived at the camp with Shelley Hack, a fashion model and one of the stars of the television show *Charlie's Angels* who was to act as the episode's guest host, he and Smith had joked about the money they could raise if Hack would pose naked with a tranquilized tiger. They were also aware that not all publicity was good publicity. Mishra recalled that the ABC director, John Wilcox, had declined to film the ritual animal sacrifices conducted before darting operations because of concerns about potential backlash from squeamish American viewers. That such concerns were not entirely unfounded is supported by the letters that both Train and Mishra received in the fall of 1980 from animal protectionists seeking to put an end to the use of buffalo calves as live bait for Chitwan's tigers. In October 1980, Mishra told a staff member of the International Society for the Protection of Animals in London that the killing of buffalo by tigers was virtually instantaneous and that, contrary to what the staffer had apparently heard, it was no "blood sport, or spectacular tourist gimmick." Nepal's parks and wildlife office, Mishra added, would welcome the society's technical or financial assistance in seeking alternatives to baiting.[51]

Ultimately the loudest complaints about the *American Sportsman* episode came not from animal protectionists but from Nepalese expatriates concerned about American scientific imperialism. In June 1981, after the episode aired in the United States, the chief of Nepal's National Parks and Wildlife Conservation Office, Biswa Upreti, sent an irate letter to the American embassy in Kathmandu. Although Upreti himself had not yet seen the film, Nepalese viewers in the United States had told him that its focus on "*expatriate persons*" such as Smith and McDougal reflected "*a bit of colonialism.*" Wilcox, the ABC filmmaker, eventually apologized to Upreti for errors in the on-screen text, including the identification of McDougal as a "director of wildlife" in Royal Chitwan National Park, but he rejected the accusation that the film had slighted the contributions of Nepalese researchers. On the contrary, he had been careful to show "the numerous Nepalese in the field; Nepalese tracking the tiger with a radio device; Nepalese working extensively with Dave Smith once the tiger was darted; and references made by the host to Hemanta Mishra regarding the number of well-trained Nepalese that were involved in the project." To the extent that Smith and McDougal had indeed been the central focus, it was merely, Wilcox argued, in order to contrast "the two methods being used in the park to identify the tigers"—namely, radio collars and photographs of facial stripes.[52]

Smithsonian administrators in Washington were pleased that Smith and Mishra had placated the tiger project's critics and impressed most of its visitors, but they were also becoming increasingly concerned that their "extracurricular" activities were detracting from their research. In February 1978, Theodore Reed, the director of the Smithsonian's National Zoological Park, cautioned Smith not to pursue a collaboration with Billy Arjan Singh, an Indian conservationist who ran a private tiger reserve in northern India just across the border from Nepal's Royal Bardia National Park. Smith had told Simons that he thought Singh's willingness to allow the radio collaring of tigers on his reserve, which was beyond the control of the Indian Forest Service, would provide an opening for the use of the technique in India, but Reed warned him that he was "skating on pretty thin ice there. Our relationships with India are made on different levels than either you or I. . . . We are the soldiers in the field but the strategy is planned elsewhere." Acceding to Nepalese government requests for help in capturing rhinos or elephants was also risky, and however necessary it might seem in order to maintain good relations, Reed advised, "your job is to get your degree." Reed similarly pressured Mishra to focus on his research and his degree rather than trying to demonstrate the project's utility to the Nepalese government. After Mishra visited Washington early in 1978, Reed warned his adviser at the University of Edinburgh, a young wildlife ecologist named Iain Taylor, that the Smithsonian's support would not last indefinitely. It was vital that Mishra complete his degree, not only for the sake of his own career but also "for the benefit of the Nation" of Nepal, which was in desperate need of trained biologists.[53]

Despite pressure from Washington to focus on their research, Smith and

For reasons of safety and convenience, much of the tracking of tigers in Chitwan was carried out on elephant-back. (Courtesy of the Smithsonian Institution Archives)

Mishra found it difficult to ignore the Nepalese government's requests for aid. Many of those requests originated in conflicts between wildlife conservation and the livelihoods of villagers in the heavily populated agricultural area around Chitwan. In 1974, in a report on nature conservation in Nepal, Mishra had noted that the country's National Parks and Wildlife Conservation Office had placed a high priority on "seeking means by which the relatively poor local people can obtain sustaining benefit from the protected areas." He and Prince Gyanendra, he later recalled, devised a three-pronged strategy to achieve that goal: providing income to local people through tourism, allowing certain renewable resources to be harvested in the parks, and removing problem animals such as man-eating tigers. As Mishra soon recognized, however, few villagers saw any benefits from Tiger Tops or other tourism operations, and the opportunity to harvest thatch grass within the park for a few weeks each year, while an important gesture, could not compensate for the loss of a single buffalo to a tiger, let alone a family member. By 1980 there were more than 260,000 people living on the borders of the park, almost double the number that had lived there a decade earlier. Though the tigers in the park rarely ventured far into the surrounding agricultural fields, a single tiger killing livestock near the park border, especially a female trying to feed her cubs, could have a significant impact. In 1977, for example, one tigress was reported to have killed eleven cattle and to have attacked a man who had come looking for his missing livestock.[54]

While the occasional loss of a buffalo might be dismissed as an acceptable cost of preserving tigers from extinction, the killing of a human demanded an immediate government response. Near dawn on November 27, 1978, a radio-collared tiger known to the Smithsonian project as 119 killed a young school-teacher who was on the way to his morning ablutions in the Rapti River at the northern border of Chitwan. Tiger 119, who had first been collared the previous January, was a subadult male with no fixed territory who was known to have had several fights with other male tigers; over the summer, tiger project staff had briefly recaptured him to treat a badly wounded leg. Within twenty-four hours of the death of the schoolteacher, the Smithsonian team had located him using his radio collar, darted him, and loaded him in a vehicle for transport to the zoo-logical park in Kathmandu. Tigers who killed humans repeatedly were punished more severely. In the fall of 1980, after a single female tiger killed three people near the park border, rumors began spreading among the villagers that the park authorities were intentionally maintaining man-eaters in order to keep villagers and their livestock out of the park. To quell what Mishra later described as a "small riot" and to prove that "although tigers are protected, the authorities also prize human life," the project staff captured the tiger and gave her a fatal drug overdose in front of hundreds of villagers.[55]

Although Mishra and Smith worked closely and successfully together, their collaboration was not without its tensions. Even before Mishra's arrival in Chit-wan as a full-time investigator in the summer of 1978, Smith had found him "pushy," as he told Simons early in 1978, though also "reasonable and very sensitive" and by far the easiest of the Nepalese officials he had encountered to work with. During Mishra's nearly year-long absence in Edinburgh, Smith had established patterns of operation that were largely successful but which did not adapt well to the addition of a second principal investigator. By the fall of 1978, frustrated with what he saw as Smith's profound lack of organization, Mishra demanded that they draw up a "gentlemen's agreement" to regularize access to equipment, staff, and money. The agreement failed to improve the situation, however, and the next May Mishra complained to Simons in a let-ter copied to Smith that "despite our personal friendship, I do not hesitate to inform you that Dave neither has experience, skill nor seems to want to admin-ister or manage the project in an orderly manner." The project's equipment was not being properly maintained, Mishra claimed, and the morale of the shikaris and other employees had plummeted so far that they had drawn up a list of formal complaints earlier that year. Unless Smith agreed to adhere to certain organizational procedures, Mishra told Simons, he saw only two alternatives: either the Smithsonian project would have to be divided into two halves with entirely separate staff and equipment, or Mishra would have to "take a more 'dictatorial stand' . . . and literally tell him what to do as he was told to do in the past before I joined the project." Counseling patience and tolerance from Washington, Simons managed to defuse the conflict without choosing either of these unpalatable alternatives, but the perhaps inevitable tension between the two project leaders continued.[56]

The relationship between the Smithsonian tiger project and India's Project Tiger improved significantly in 1979. In February 1979, Smith, Mishra, and McDougal attended a symposium organized by Project Tiger. Smith later described the symposium as the crucial turning point after which genuine collaboration between tiger researchers in India and Nepal became possible for the first time. Ripley, who also attended, was surprised to find the Smithsonian team was welcomed by the Indian scientists and reserve managers; he later told British conservationist Guy Mountfort that "it was a revelation to me to note the sincere and rapt attention given to the Nepal party when they came on the platform and delivered their reports." Both Smith's and McDougal's presentations focused on the dynamics of Chitwan's expanding population of highly territorial animals. McDougal told the audience that "even under optimum conditions, there is a limit to which territories can be compressed, and hence to the population of reproductive adults," which meant that the Nepalese government would eventually need to protect more land if it wanted a larger tiger population.[57]

Since 1975, when the Smithsonian had first begun contributing its own funds toward research in Chitwan, WWF-US had continued to provide about half of the project's funding. In the late 1970s, WWF-US's priorities began to change under the leadership of its first paid president, Russell Train, who was elected by WWF-US's board in the spring of 1978. A former federal tax court judge who had shifted his focus to conservation after going on several safaris in East Africa in the late 1950s, Train had served as the first, unpaid president of WWF-US but had been forced to resign from the organization in 1969 when Richard Nixon's administration chose him for the position of undersecretary of the Department of the Interior. Train's term at Interior was followed by three years as chairman of the Council on Environmental Quality and four

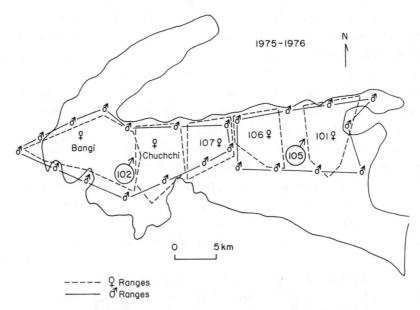

Radio tracking helped biologists determine the territories of individual tigers, as in this map by Melvin E. Sunquist of the exclusive ranges of male and female tigers within the borders of Royal Chitwan National Park. (Source: Melvin E. Sunquist, *The Social Organization of Tigers* [Panthera tigris] *in Royal Chitawan National Park, Nepal* [Washington, DC: Smithsonian Institution Press, 1981], 48, figure 27, courtesy of the Smithsonian Institution Scholarly Press)

years as administrator of the Environmental Protection Agency. Ripley played a key role in convincing Train to take the WWF-US position, arguing that the organization had the potential to grow beyond its small and amateurish beginnings. When he accepted the position, Train later recalled, he "was determined to see our program grow in terms of money, geographic scope, and, above all, in conservation effectiveness." One of his main strategies for doing so was to bolster WWF's own conservation programs. As long as WWF acted primarily as a grant-making organizing, Train believed, it would have difficulty raising funds from philanthropic foundations and other major donors, who had little interest in providing money that would simply be passed along to other organizations after WWF-US had taken an administrative cut. Train also tried to shift WWF-US toward what he later described as a "broader ecological approach" and away from "its early emphasis on species conservation."[58]

Train's new vision for WWF-US meant that even as the leadership of IUCN and WWF International seemed finally to be accepting the legitimacy of the

Chitwan project, WWF-US was looking for ways to bring its involvement to a graceful conclusion. The Smithsonian, too, was preparing for a significant change in the nature of its operations in Nepal. When Challinor sent the Smithsonian's request for another year of funding to Train in the summer of 1979, he emphasized that the tiger project was beginning to pay off in concrete conservation outcomes. The "results of the project to date have already influenced His Majesty's Government to twice extend the size of Chitwan National Park," he wrote, and future work by Smith and Mishra would inform land management plans for tigers throughout Nepal. But Challinor and other Smithsonian administrators were already becoming convinced that they would never be able to conduct truly ecological research in Chitwan as long as tigers remained the central focus. Troth's eviction had only been the first and most troubling indication of the limited extent to which ecological research could be justified "under the tiger rubric." Early in 1978, Simons had complained to Ripley about the government's "vexsome" lack of appreciation for "overall ecosystem work." In the spring of 1979 Wemmer told Simons that he also had found that the "meaning of wildlife conservation in Nepal hasn't advanced much beyond the concept of big game management as it was understood in the days of the British raj." There were no real naturalists among the Nepalese officials, Wemmer added, with the important exception of Mishra, and "as long as the tiger has center stage, everything of indirect relevance to the tiger is felt by the Nepalis to have *no* relevance."[59]

Without losing its focus on tigers, the Nepalese government also had begun to consider the desirability of significant changes to the Smithsonian's activities in Chitwan. In the fall of 1979, Chitwan's largest male tiger collapsed and drowned in a small pool, having bolted into dense vegetation before the immobilizing drug with which he had just been darted had taken effect. In response to the death, the chief of the National Parks and Wildlife Conservation Office, Biswa Upreti, told Simons that he was placing an indefinite hold on further darting. The Nepalese government considered the death a "purely accidental case, but in the mean time, we are also forced to think about the justification of continuing immobilisation of the tiger since about 50% of the tigers in Chitwan have already been radiocollared." Smith had been instructed to continue tracking tigers that had already been collared but to stop further darting until the Nepalese government had had an opportunity to review the project's findings. To that end, Upreti asked Simons to send him the final reports of all previous Smithsonian-sponsored researchers. A few months later, in a memo on the

Smithsonian's changing plans in Nepal, Heck noted that Prince Gyanendra had agreed that requests to dart additional animals would be considered on a case-by-case basis, but the general moratorium would remain in place.[60]

A prohibition on darting would have been disastrous earlier in the project's history, but by early 1980 the Smithsonian had already made the decision to bring its tiger research to a close once Smith and Mishra had finished the fieldwork necessary for their doctoral degrees. Instead of a permanent research station on the model of the Smithsonian Tropical Research Institute in the Panama Canal Zone, Challinor now argued, some sort of conservation trust would be the best institutional mechanism for facilitating Smithsonian research in Nepal. In February 1980, he told Prince Gyanendra that there were a number of successful examples of such trusts in other countries, including the Bahamas National Trust, the Falkland Islands Foundation, the Seychelles Islands Foundation, the Darwin Foundation in the Galápagos Islands, and the Serengeti Research Institute. He also sought to convince Train that WWF-US could use such a trust to further its own ends in Nepal. Train later recalled that the Smithsonian, having decided that it was finished with tiger research, began encouraging WWF-US to become involved in "wildlife conservation more generally" in Nepal and to support the creation of the trust. While such a trust would probably depend heavily on grants from WWF-US, especially at the beginning, it would also be able to raise its own funds directly, and it would provide a local nongovernmental partner for the new conservation programs that Train hoped to establish.[61]

In August 1980, Mishra received official confirmation that WWF-US was ending its support for the Smithsonian-Nepal Tiger Ecology Project. The decision would be a major blow to Nepal's wildlife conservation and management efforts, he told WWF-US's Thomas Lovejoy, since the project had effectively become the "scientific wing" of Nepal's parks and wildlife department. Without the support of the Smithsonian, such arguments had little chance of success. With the loss of WWF funding, research activities in Chitwan were significantly reduced, and most of the staff were let go. Smith had left Chitwan in the summer of 1980 to begin writing up his thesis, and Mishra remained in the park into early 1981 to complete his own thesis research on chital deer, one of the tigers' primary prey species. After Smith's departure, radio tracking of the tigers whose collars were still operational continued, with McDougal covering the tigers at the west end of the park while Mishra monitored those at the east end of the park. In the spring of 1982, former chief shikari Prem Bahadur Rai told Wemmer he was trying to start a small hotel and wildlife tourism operation at Sauraha.[62]

Tales of the Terai

Train and his wife, Aileen Train, visited Chitwan in early 1981, while Mishra was still wrapping up his fieldwork. Before leaving Washington, as Train later explained to the American ambassador to Nepal, he had spoken with Challinor about the proposed conservation trust, and he hoped to convince Prince Gyanendra of its value during his visit. On their way to the Smithsonian field camp, the Trains stopped at the park visitors' center, where they noticed a photograph of King George V, whose hunting party had killed thirty-eight tigers and eighteen rhinos in Chitwan seventy years earlier in the course of an elaborate hunt involving hundreds of elephants. In his journal, Train noted the strange continuity between such colonial-era hunts and the Smithsonian project's research: "While we can be glad that such massacres are a thing of the past, a sad aspect of the matter is that the training and keeping of elephants is a rapidly disappearing art. Our tiger project is one of the last—perhaps even the last—examples of their utilization in this fashion."[63]

After lunch with Mishra and his wife at the tiger project's camp, the Trains were invited to participate in an attempt to radio-collar the cub of a previously collared female leopard. Mishra later recalled that he had been asked by Lovejoy to show the Trains "the field techniques developed in Nepal, particularly how we use old Nepalese hunting techniques with modern scientific tools to catch large mammals." As usual on days when darting operations were to be carried out, Mishra and the project's shikaris and elephant drivers had gathered earlier that day at the base of a silk cotton tree—the same *Bombax ceiba* that Troth had come to Chitwan to study—for a ceremony to bless the effort. Beneath what Train described in his journal as "bright, orange red, trumpet-shaped blossoms" of the silk cotton trees, the chief elephant driver, Badai Tharu, performed a puja for the goddess of the forest. As his surname indicated, Badai was a member of the Tharu tribe that had dominated the Chitwan area until the eradication of malaria and subsequent large-scale immigration from the Himalayan foothills in the 1950s and 1960s. Two pigeons, a chicken, and a goat were sacrificed, their blood was sprinkled on the ground, and tikka, a mixture of rice and red dye, was dabbed on the foreheads of the participants.[64]

In midafternoon the Trains joined the team on elephant-back to search for the leopard cub. After Prem Bahadur Rai had located the mother with the help of her radio collar, Mishra stepped off his elephant into a nearby tree to wait with a dart gun. A white cloth several hundred yards long and about three feet high was unrolled to make a V-shaped funnel ending at Mishra's hiding place, and the four elephants swept in a line through the area where the leopard and her cub were thought to be hiding. After several sweeps the collared leopard burst out of hiding but without her cub, so Mishra called off the hunt and the team returned to camp for dinner and drinks. That night the Trains could hear, as Russell Train wrote in his journal, "a clamor raised from the villages a half mile away as the natives chased marauding animals from their fields." The next morning they went out with the team again to attempt to capture and radio-collar a large male leopard, but this also failed, to no great displeasure on the part of the Trains. Train noted in his journal that he and his wife "both agreed that we were rather glad that the leopard—a really magnificent animal and the biggest leopard we had ever seen—had escaped without a collar."[65]

After several years of negotiation among the Smithsonian, WWF-US, and the Nepalese government, the King Mahendra Trust for Nature Conservation was formally established in 1983, with Mishra as its first director and with the Smithsonian and WWF-US as key financial supporters. In the fall of 1983, after meeting with the director of the U.S. Agency for International Development for Nepal, Simons noted that the Smithsonian "has been and probably will be the principal and continuous source of foreign assistance to HMG's Department of National Parks and Wildlife Conservation." While Mishra had day-to-day control of the King Mahendra Trust, its policies and fundraising efforts were coordinated by a board of trustees that included representatives of the Nepalese government as well as representatives from the United States and the United Kingdom. Initially the two non-Nepalese trustees were Train, the president of WWF-US, and Sir Arthur "Jerry" Norman, chairman of the board of WWF-UK. Train later recalled that there was an implicit "special relationship" between the King Mahendra Trust and WWF-US: in exchange for major financial support, the trust would facilitate WWF's new conservation initiatives in Nepal.[66]

In a May 1984 speech about the new trust given during a visit to the Smithsonian in Washington, Prince Gyanendra noted that the Smithsonian-sponsored research in Chitwan had featured "the combination and use of indigenous, traditional, as well as modern methods of fauna observation including the use of radio telemetry monitoring," a kind of cultural cross-fertilization that he expected to continue under the trust's guidance. Although the tiger project had come to an end, the Smithsonian had remained closely involved in research in Chitwan. It had provided a fellowship to Sanat Dhungel, one of the assistants on the tiger project, to conduct a study of hog deer for his doctorate, and it funded a radio-tracking study of the gharial, an endangered crocodile living in the rivers around Chitwan. After meeting with Ripley in Washington, Prince Gyanendra agreed to a name change that officials in the National Parks and Wildlife Conservation Office had been loath to grant: the "Tiger" in the Smithsonian-Nepal Tiger Ecology Project would now be replaced by "Terai" to reflect the project's landscape-level focus.[67]

Eric Dinerstein was the first American field biologist to work in Chitwan under this broadened mandate, while Mishra served as project coordinator from the King Mahendra Trust office in Kathmandu. As a Peace Corps volunteer in Nepal in the 1970s, Dinerstein had attempted to census the tigers of Royal Bardia National Park at Mishra's request. That experience had given Dinerstein an unusual level of familiarity with the landscape and conservation issues of the Nepal Terai when he returned in 1984. He was also a fluent speaker of Nepali, a first among the Smithsonian biologists in Chitwan. Mishra later claimed that Dinerstein "could easily be mistaken for an upper-class Nepali," and Wemmer predicted in March of 1984 that his cultural familiarity and language skills would be a "great benefit to the program." Dinerstein's and Mishra's acculturation to Nepal and the West, respectively, were the subject of constant joking in their correspondence with Smithsonian administrators, as in a letter Wemmer sent later that year to "Eric Bahadur Dinerstein" and "Hemanta Erasmus Mishra."[68]

Dinerstein's research was intended to embody the Smithsonian-Nepal Terai Ecology Project's broadened mandate by examining how the highly endangered one-horned rhinoceroses distributed plant seeds across the landscape. By the summer of 1985, however, Mishra was becoming skeptical about the value of the project, which he believed had been diverted by the glamorous work of radio-tagging and tracking rhinos from its original goal of understanding a large herbivore's impact on plant distribution. Moreover, the working relationship between Mishra and Dinerstein had become strained. In August 1985 Dinerstein told Wemmer that he had "the sneaking suspicion that somewhere along the line I am going to get screwed by Hemanta," which he thought was "not a unique sentiment among white biologists who have worked in Nepal." A few weeks later, he recommended a recent article in the *Bulletin of the Ecological Society of America* to Wemmer on the subject of "how us'ns have offended they'suns and some criteria for judging how sensitive (or how ugly) we biologists really appear when viewed under eastern eyes." Compared to the projects and problems described in the article, Dinerstein thought, the Smithsonian's work in Nepal "comes up smelling like roses."[69]

Mishra's assessment of the Smithsonian's record in Nepal was less charitable. Smith, he told Simons, had left Nepal just as he was finally beginning to understand "the problems, issues and politics of wildlife conservation in Nepal," and Dinerstein, whose current research seemed more likely to help "his ego and image" than to assist conservation and management in Nepal, was likely to do the same. While Mishra had long defended the value of the Chitwan research to Nepalese bureaucrats who might otherwise have been convinced by the "idiots" at IUCN that the Smithsonian was engaged in "scientific imperialism," he told Simons, it was time for the approach to change. Seemingly minor conflicts, such as Mishra and Dinerstein's inability to devise a satisfactory scheme for sharing the project's only personal computer, reflected a deeper, long-standing problem. Each of the American biologists in Chitwan had had his technological "hobby horse," Mishra told Simons; for Smith it had been airplanes, for Dinerstein it was computers, and "the next one, I am sure will be into satellites or video machines. Any way, the circle will be repeated and life will go on my friend with you and me exchanging similar letters untill we cross the age one half of a century. Then, it will be time to retire and jointly write the stories or expose behind the real-politik and bullshit about the 'Tales of the Terai,' with a hope that it will bring us both fame and fortune that was not possible as we were envolved in doing all the dirty works and getting things done 'to promote science and save Nepal's endangered species,' which, by the way, would be gone by them as we Nepalis keep on breeding at a fast rate as as at present."[70]

Despite the bitterness of their letters to Washington, Mishra and Dinerstein soon patched up their differences, with some help from Wemmer's and Simons's cooler heads. In November 1985, after Mishra had gone out of his way to support Dinerstein's project during an important meeting with visiting WWF-US representatives, Dinerstein told Wemmer that his earlier fears of betrayal had been misplaced: "I officially eat crow." Several decades later, when both Dinerstein and Mishra had published their "Tales of the Terai," Dinerstein would describe Mishra as "perhaps the most innovative Asian conservationist of my lifetime," while Mishra would describe Dinerstein as a "brilliant scientist" with an "in-depth knowledge of the language and culture" of Nepal.[71]

Like Smith and Mishra before him, Dinerstein spent a significant amount of his time in Chitwan demonstrating the project's techniques to visiting VIPs. Though tigers were no longer radio-collared after 1980, rhinos were darted on fifty-one occasions between 1985 and 1988; twenty-one different rhinos were radio-collared, and some were translocated to other reserves in an effort devised by Mishra and supported by the Smithsonian, WWF, and the Nepalese government to increase the numbers and resilience of the population. Except for the absence of baiting, the procedure was similar to that used with tigers; as Dinerstein later explained, it "depended heavily on the skills of trained elephants and their drivers." As in the days of the tiger project, the darting operations were the highlight of visits to the research camp. In 1986, the Smithsonian project welcomed Prince Philip, the Duke of Edinburgh, who had previously visited Chitwan in 1961 for what had been predicted to be "the tiger shoot of the century" but had turned into a much more modest affair after the British press pointed out the apparent contradiction between the tiger hunting and Philip's role as president of the newly established WWF-UK. On his return to Chitwan in 1986, as historian John MacKenzie recounts, the nearest that Philip came to "big-game shooting was to witness the tranquillising of a rhino called 'Philip' so that it could be fitted with a radio device." Train too had a rhinoceros named in his honor when he returned to Chitwan in 1987 and participated in a darting effort led by Mishra. In his journal, Train described the method of driving the rhino so that it could be darted as "an imposing sight—reminiscent of the old days of royal hunts for tiger and other quarry—as our 21 elephants moved slowly 'line-abreast' through the shoulder-high grass of the river plain."[72]

In the late 1980s the Smithsonian ended its support for the Smithsonian-Nepal Terai Ecology Project. Dinerstein left the project in 1988 to become WWF-US's chief scientist, in which position he would return frequently to Chitwan over the next several decades. In the summer of 1989, Mishra tried to convince the Smithsonian to support a new "Nepal Conservation Training and Research Institute" at Sauraha, but Simons told him that the Smithsonian's "long-range research plans in Asia are shifting." The Smithsonian might be able to support research in Nepal on a project-by-project basis, but it was "no longer in a position to support the camp, or any form of maintenance of a larger entity," and funding for the Sauraha camp would cease that fall. The Smithsonian looked forward, Simons told Mishra, to seeing the King Mahendra Trust and other projects "flourish under Nepalese leadership." In the spring of 1990, Nepal was swept by protests against the monarchy that eventually led to a series of partial democratic reforms. Prince Gyanendra wrote to Train to reassure him that "business is going on as usual at the King Mahendra Trust for Nature Conservation, notwithstanding these political developments," but over the summer Train heard rumors from Norman that the recently renamed Department of National Parks and Wildlife Conservation was trying to take over the King Mahendra Trust, from which Mishra had recently been on "extended leave" while thinking about launching an environmental consulting company. Train told Norman he shared his concerns about the future of the trust, noting that "these are troublesome times and I imagine they are going to get worse!" The King Mahendra Trust, like the monarchy, survived the political upheaval of 1990, but within a few years Mishra had left Nepal for a job at the World Bank, while Chitwan had become what Mishra later described as "the hot spot of a bloody rebellion against the monarchy in Nepal."[73]

Over the course of the 1980s, Indian tiger conservationists had become increasingly open to the use of the techniques pioneered in Nepal by the Smithsonian researchers. In 1980, Project Tiger had invited Mech and Seal to lead a second workshop in India on tranquilizing and telemetry, during which they were allowed to immobilize an Indian tiger for the first time. Indian biologists also began to receiving training in these techniques through the Smithsonian's Conservation and Research Center in Front Royal, Virginia. In 1981, Rasanayagam "Rudy" Rudran, a Sri Lankan who had come to the United States in 1970 for his doctorate, began offering a wildlife management and research training program at the CRC. One of his first students was the Indian biologist A. J. T. Johnsingh, who learned to radio-track raccoons and opossums for his postdoctoral research at the Smithsonian before returning to India in the fall of 1981, where he subsequently introduced the technique to his own students. In 1982, the Indian Forest Service invited fifteen American wildlife biologists to meet with Indian wildlife managers and scientists at Kanha National Park, the site of Schaller's research for *The Deer and the Tiger*, where they discussed the adaptation of scientific wildlife management to the conditions of South Asia. Later that year, the Indian government established the Wildlife Institute of India, one of whose main goals was to promote the use of "modern technologies" in Indian wildlife biology.[74]

By the late 1980s, according to K. Ullas Karanth, even Sankhala had become more open to the techniques of "modern tiger research." Karanth was a former engineer and farmer with a strong interest in tiger conservation in southern India who had met Sunquist, Seidensticker, and Smith at an international wild-life conference sponsored by the Bombay Natural History Society in 1983. They in turn introduced him to Rudran, whose wildlife management and research training program he attended in the summer of 1984. From there Karanth moved to the University of Florida, where he studied for a master's degree under Sunquist, who continued to advise him after he returned to India to earn his doctorate.[75]

In January 1990, Karanth began the first radio-tracking study of tigers in India, capturing and collaring four tigers in Nagarhole National Park using techniques modeled directly on those first used in Chitwan fifteen years earlier, including the use of live baits, white cloth fences, trained elephants, and dart guns. In a later account of the research, Karanth praised "the ingenuity of Nep-alese and Bihari shikaris who invented this technique, once employed by roy-alty to slaughter tigers," while also noting the irony of transforming a hunting technique into "one of the scientific tools I use to help preserve these big cats." Although the darting and collaring of the Nagarhole tigers took place without incident, five dead tigers were discovered in the park later that spring, includ-ing one collared male, and Karanth's research was accused of being responsible for the deaths. While an investigation into the deaths proceeded, Karanth was temporarily excluded from working in the park. One of Karanth's colleagues, Raman Sukumar, came to his defense in an article that noted that similar tech-niques had been used far more extensively in Nepal with only two confirmed deaths caused by darting and that, in any case, there was no evidence linking the deaths in Nagarhole to Karanth's research. An unfounded prejudice against the technique, Sukumar lamented, threatened "to shut the doors once again to use of such technology by Indian biologists."[76]

Karanth was eventually allowed to return to Nagarhole and to continue mon-itoring the tigers he had already radio-collared, but instead of collaring addi-tional tigers, he focused on developing less invasive but still rigorous methods of censusing the park's tigers. The most successful of these was the use of motion-triggered cameras, a technique that McDougal had begun experimenting with in Chitwan in the late 1970s. Karanth and his team placed paired cameras along trails in Nagarhole and other tiger reserves, which were triggered when a pass-ing animal interrupted an infrared beam. The resulting photographs of stripe patterns on both sides of a tiger could be used to identify individuals far more reliably, Karanth argued, than could pugmark tracing. Karanth also worked with James D. Nicholls, a statistician at the Patuxent Wildlife Research Center in Maryland, to show how each photograph of a tiger could be modeled as a "capture" of that particular animal, thereby allowing familiar capture-recapture techniques in wildlife biology to be used to estimate the size of the population. Used this way, Karanth and Nicholls argued in 1998, camera trapping was a more promising census technique than either pugmark tracking or radiotelemetry. The former was based on what they called "demonstrably erroneous" assumptions about the distinctiveness of tracks, while the latter was "constrained by the small number of animals that can be tagged simultaneously, uncertainty about numbers of untagged tigers, and the high cost and effort involved." When Fiona Sunquist had interviewed Karanth for an article on camera trapping published in *International Wildlife* the year before, he had mentioned another advantage to the technique: "You don't have to catch and handle the animal."[77]

By the late 1990s, radio tracking of big cats was being overshadowed by new techniques that did not involve capturing animals and were better suited to exactly the kinds of countrywide surveys defended by critics of Hornocker's and Seidensticker's initial tiger research proposal in the early 1970s. In 2002, in a summary of recent advances in research techniques for studying cats, Mel and Fiona Sunquist noted that the introduction of radiotelemetry in the 1970s had "turned a spotlight on the previously hidden lives of the secretive, solitary felids" and was now "an integral part of most cat studies." Still, they argued, conventional radiotelemetry had a number of drawbacks, including high cost, effort, and risk and the fact that radio signals "can be reflected, deflected, or obstructed," which made it problematic for studying species that lived in mountainous terrain or in habitats with dense vegetation. Satellite collars equipped with Global Positioning System receivers might obviate some of these limitations, but the danger of collaring remained. "No matter how many individuals you handle," the Sunquists wrote, "the process of trapping and tranquilizing an animal in the field is always fraught with anxiety. Even with the greatest care and expert attention, animals occasionally injure themselves in traps and snares, or die under anesthetic." Fortunately, new techniques such as camera trapping and the analysis of DNA in feces and hair could "make life easier for the animal and are often better received by landowners and wildlife departments." These "new and largely noninvasive techniques," the Sunquists suggested, promised to be as revolutionary for the study of tigers and other species as radiotelemetry had been a quarter century earlier.[78]

CHAPTER FOUR

The Regulatory Leviathan

While radio tags had been successfully deployed on a variety of species of birds and terrestrial animals by the mid-1960s, the use of the technique to study marine animals lagged far behind. The technical and logistical challenges of radio-tagging marine mammals, sea turtles, and fishes were daunting: the animals often lived in harsh and inaccessible environments; they were difficult to locate, capture, and handle; and the fact that water blocked radio signals made it infeasible to track any species except those that spent significant amounts of time at or near the water's surface. One of the first researchers to explore the possibility of radio-tracking marine animals was William E. Evans, a bioacoustics expert whose work at Lockheed Aircraft Corporation on U.S. Navy–sponsored projects to map the underwater soundscape had sparked his interest in technologies that would make it possible to consistently relocate individual dolphins. In 1962, Evens attended the Office of Naval Research–sponsored conference on biotelemetry at the American Museum of Natural History, where he and Lockheed colleague W. W. Sutherland presented a proposal that they had developed to radio-tag dolphins. Though they were enthusiastic about the possibilities opened up by the technique, Evans and Sutherland thought studies of captive animals were just barely feasible given the state of the technology, while field studies of free-swimming dolphins were "hardly practical. The attachment of a telemetry package would involve either capture and release of the animal or some other means such as the use of a harpoon. In either case if the observer didn't lose the equipment and/or the animal, the manipulation involved could conceivably contaminate the data collected." Perhaps, they suggested, a remotely controlled craft that relayed the signal from the dolphin's tag back to a mother ship would reduce the effect of the observers on the animal once he or she had been tagged, but they had "no illusions to the fact that we are going to get data that are going to be unaffected by any techniques that we use."[1]

Though Evans and Sutherland were the first to publicly propose radio-tracking marine mammals, two other ONR-supported bioacousticians, William Schevill and William Watkins, were the first to report field tests of the technique. Schevill and Watkins were based at the Woods Hole Oceanographic Institution, where they studied the vocalizations of North Atlantic right whales and other species of cetaceans. Like Evans and most other American biologists studying whales and dolphins at the time, the majority of whom were funded by ONR, Schevill and Watkins were primarily interested in questions of information and communication. Radio tracking, they hoped, would make it possible to monitor the relationship between vocalizations and behavior in individual cetaceans over extended periods of time. In the fall of 1964, WHOI hosted a conference on the oceanographic applications of high-flying aircraft and artificial satellites, which held out the possibility that, as conference organizer Paul Fye put it, for "the first time the whole world ocean could be placed under continuous surveillance for purely scientific purposes." In the paper he gave at the conference on the radio tracking of whales, Schevill noted the technical challenges of devising a waterproof, hydrodynamic tag and a ship-based tracking system as well as "the sporting uncertainty of the actual tagging of the whale." Unlike the small cetaceans that Evans and Sutherland had hoped to tag or the terrestrial and avian species with which most researchers worked, adult right whales were far too large to be captured or restrained. By the following year, Schevill and Watkins had developed a small, waterproof cylindrical radio tag with an antenna at one end and a barbed metal dart at the other, which they attached to the end of a weighted pole that could be dropped from a helicopter hovering over a whale as it surfaced. Although they managed to attach a few tags using this risky method, the tags were often damaged in the attachment processes. Even in the handful of cases in which the researchers were able to detect a signal from a surfacing whale, the whale inevitably submerged again before they were able to triangulate its exact location.[2]

Most researchers who attempted to radio-track whales, sea turtles, and other marine animals in the 1960s experienced similar frustrations. In 1965, Lowell Adams, one of the founding members of the Wildlife Society's Telemetry Committee, proposed radio-tagging gray whales at their breeding grounds in the lagoons of Baja California to the National Science Foundation, but his proposal was rejected on the grounds that it was technologically infeasible. Carl L. Hubbs, a marine biologist at the Scripps Institution of Oceanography, told one colleague he also opposed Adams's experiment because "the chances of much scientific information resulting is minimal" and because "excess disturbance of the whales in the Lagoon may seriously affect their mass reproductive operations." When Adams contacted him for advice on working in the lagoons, Hubbs reminded him that, since gray whales were a protected species, he would need a permit from the Mexican government if he planned on "tagging the whales or otherwise seriously disturbing them." Although Adams managed to find a private donor to support a trip to Baja California, his whale-tagging efforts were both short lived and unsuccessful.[3]

Adams had never studied cetaceans before proposing to tag gray whales, but even better-qualified researchers found whale tracking extraordinarily difficult. The cetologist Roger Payne tried unsuccessfully to track a humpback whale in 1967 using an acoustic tag, which emitted a tone that could be detected underwater, and he repeatedly submitted proposals for whale-tagging studies to the National Science Foundation over the next several years. Despite Payne's qualifications, his proposals were rejected as both technically unfeasible and likely to disturb the animals' sensitive acoustic systems. As late as 1970, when an engineer named Jack Schultz proposed a telemetry study of blue whale migration, the main reason NSF gave for rejecting the proposal was that it was technologically premature. The director of NSF's Office of Polar Programs, George Llano, told Schultz that a panel of twelve specialists had recently reviewed and rejected a similar proposal: "While, like you, they all endorsed the need for developing a system for whale tracking, to a man they stated that current results are based on unconfirmed techniques resulting in questionable data." NSF's hesitation to invest in radio tracking seemed further justified, Llano added, by the fact that ONR, NASA, and the Smithsonian Institution were "still struggling with" the development of the technique.[4]

Evans was the first to prove that radiotelemetry was a viable technique for studying cetaceans. Since presenting his proposal with Sutherland in 1962, he had studied with Kenneth Norris, a marine biologist at the University of California, Los Angeles, and one of the cofounders of Sea World, and he had begun working at the navy's Marine Mammal Laboratory at Point Mugu, California. Although Norris shared Evans's interest in biotelemetry, it was not until 1967 that Evans began working in earnest on the technique. Between 1967 and 1971, Evans collaborated with Ocean Applied Research, a firm near San Diego that built radio beacons and automatic direction finders for oceanographic buoys, to develop a small cetacean radio tag and tracking system. Like Payne, Evans initially experimented with acoustic tags because they could be detected even when they were submerged but eventually ruled them out because of concerns about the animals' dependence on sound for communication and navigation. Instead he repackaged one of OAR's submersible radio beacons in a hydrodynamic package that could be bolted to a dolphin's dorsal fin, using one of the performing dolphins at the Marineland of the Pacific theme park near Los Angeles to test the prototype. Evans's key advance over earlier efforts, however, was not in the radio tag but in the use of OAR's automatic radio direction finder, which could identify the direction of radio signals that lasted for as little as a fraction of a second. This solved the problem that Watkins and Schevill had faced of being unable to triangulate the location of an animal during the brief time that its tag was above the surface.[5]

In addition to its use in Evans's research on the movements of dolphins off the coast of southern California, Evans's radio-tracking system quickly found applications in the U.S. Navy's Marine Mammal Program. One of those applications was keeping track of killer whales and pilot whales who had been trained to retrieve objects from the sea floor at depths of hundreds or thousands of feet in an experimental project called Deep Ops, which was conducted primarily at the Naval Undersea Center's field station in Hawaii. The first two animals used in the project were killer whales named Ishmael and Ahab, who had been captured in Puget Sound by the Seattle Public Aquarium and transferred to the navy in 1968. After "basic training" at NUC headquarters in southern California, the orcas were moved to Hawaii for training in the open ocean. Although NSF's Office of Polar Programs had considered a proposal to radio-tag orcas in 1966, no one had ever attempted to do so, let alone succeeded. Concerned that Ahab, Ishmael, and the two pilot whales used for the project might try to escape once released from their sea pens, Deep Ops staff adapted Evans's radiotelemetry system so that it could be attached to the cetaceans' dorsal fins with a removable harness. The first animal to be fitted with the harness was the pilot whale Morgan, who made vigorous efforts to remove it by rubbing against the side of his pen. Even after Deep Ops personnel punished Morgan with an electric cattle prod until he stopped rubbing, he was able to dislodge it with energetic swimming, jumping, and spinning. Eventually a backpack was designed that Morgan and the other Deep Ops animals were unable to remove and seemed willing to tolerate. The radio tag proved its value in open-ocean tests with the two killer whales, both of whom were temporarily lost. Ahab was quickly relocated with the help of his radio tag, but Ishmael was lost permanently, in large part because his tag malfunctioned. Although Project Deep Ops was terminated in 1971, the radio-tracking system that Evans had developed continued to be used elsewhere in the Marine Mammal Program.[6]

It also quickly found its way into discussions about cetacean conservation. In the fall of 1970, Secretary of the Interior Walter Hickel announced that eight species of great whales would be placed on the federal endangered species list and that an international scientific conference would soon be held to determine the status of the world's whales. In June 1971, that conference was convened by Schevill, Norris, and marine mammalogist G. Carleton Ray at Shenandoah National Park. Although most of the conference papers focused on methods of estimating global whale populations, the two papers that focused on field methods were both devoted to radio tracking and featured Evans as author or coauthor. In one of the papers, Norris, Evans, and Ray argued that telemetry would allow biologists to obtain far more detailed data on whale movements than were available from extant methods of marking, such as "Discovery marks," named for the research committee that had first deployed them in the 1930s. These were thin metal tubes imprinted with serial numbers that were fired into the blubber of whales and recovered after the whales were killed and processed for oil. Norris, Evans, and Ray argued that whereas Discovery marks provided only two data points, "the point of marking" and the "point of death," radio tags had the potential to provide detailed maps of marine mammal migrations.[7]

Radio tagging would also make it possible for whale research to be logistically independent of commercial whaling, something most of the participants in the Shenandoah conference agreed was highly desirable. The conference's Working Group on Biology and Natural History concluded that the data from "commercially caught whales," upon which virtually all estimates of global whale populations were based, were incapable of guiding "biologically sound harvesting." Two techniques, the working group argued, could compensate for the deficiencies of commercial data: pelagic research captures and the "development of methods for the study of individual whales through radio-telemetry." Unlike Discovery marks, radiotelemetry would give "continuous records of individuals" and make it possible for population analysis to be "uncoupled" from commercial whaling. It would allow biologists to move past "inferences drawn from dead animals and brief observations of live animals when they were near the surface" toward a richer understanding of cetacean behavior. The working group's recommendations were reflected in the conference's summary report, which claimed that "tracking and biological studies by use of radio-telemetry, recoverable instrument packages, transponders, aircraft and satellite monitoring" had "special merit" for understanding the migrations of whales and the relationships between different populations of the same species.[8]

Although policy questions were excluded from the Shenandoah conference, the scientists who participated hoped their conclusions would influence ongoing and increasingly contentious debates over the conservation of whales and other marine mammals in Washington. Representative David Pryor, a member of the Democratic Party from Arkansas, had recently introduced House Resolution 6558, a bill to prohibit killing, injuring, or harassing marine mammals. To many marine mammalogists, even those in favor of heightened federal protection for marine mammals, the Pryor bill seemed to threaten the basic tools of conservation and wildlife biology, most of which involved some degree of manipulation of wildlife populations. In the spring of 1971, Hubbs told one member of the California congressional delegation that H.R. 6558, if passed, "would put all oceanariums out of service, cripple all major zoos, force fur sealing from sane management to destructive pelagic operations, render modern tuna fishing illegal and make a farce of sound conservation." Hubbs also recruited his fellow marine mammalogists to campaign against the bill, including one of his former graduate students, Norris. Hubbs told Norris that although the deficiencies of the bill were obvious to experts, who might therefore assume that it had no chance of passing Congress, "we see so much exuberance in conservation now days, much of it gone way out of control, that I'm sure we need to take action."[9]

Although conservationists generally supported strengthened government protection for whales, dolphins, seals, and other marine mammals, much of the public pressure for action came not from them but from animal protectionists motivated by humane concerns over the killing of baby seals for fur in Canada, the bycatch of large numbers of dolphins in tuna fisheries, and the slaughter of great whales. To protectionists such as Cleveland Amory, the president of the Fund for Animals, the provisions for "rational management" built into an alternative to Pryor's bill introduced by Representative Glenn Anderson failed to respond to public opposition to the continued massacre of sentient beings. At congressional hearings on ocean mammal protection held in the fall of 1971, Amory testified: "We are, after all, not just talking about extinction and endangered species here. . . . It does not matter how many of an animal or sea mammal there is. What really matters is the total immorality and senselessness of taking any such creatures for . . . a frivolous purpose."[10]

As historian Roderick Nash writes, the entry into debates over marine mammal protection of the idea that individual animals had a right not to be harmed regardless of the health of the population to which they belonged "surprised and dismayed utilitarian conservationists and professional wildlife managers." Strict protection contravened the philosophical commitment to the efficient and sustainable use of natural resources in which they had been trained as well as directly and indirectly threatening their ability to conduct research. At the fall 1971 hearings, Schevill, Norris, and Ray testified before the House Subcommittee on Fisheries and Wildlife Conservation in favor of Anderson's less restrictive bill, H.R. 10420. While some might find the idea of managing wild animals and their habitats distasteful, they told the committee, such management was necessary in a world where pristine, untouched ecosystems no longer existed: "We have already intervened, and often deeply, in systems involving marine mammals. Therefore, we must manage, if only to assure that this intervention is kept in control." Rational management in turn depended on science, and Schevill, Norris, and Ray urged the committee to include support for marine mammal research in whichever version of the bill they eventually passed. In their testimony and in the report of the Shenandoah conference they submitted for the record, they identified radiotelemetry as one of the research techniques most deserving of support. When one member of the House committee asked why they placed so much emphasis on this particular technique, Norris replied that existing methods had left marine mammalogists in a state of profound ignorance: "We do not know the numbers of those animals. We do not know where they go. . . . Hence, these tracking needs are great, and we are faced with a severe budget constraint in having the chance to try such tracking."[11]

Although a number of small cetaceans that could be captured and restrained had been radio-tracked by the time of Norris's testimony in the fall of 1971, albeit never for longer than three weeks, large whales continued to pose formidable challenges. In the spring of 1972, Evans succeeded in tagging and tracking a gray whale, but only because the animal was an adolescent who had been kept in captivity at Sea World during the previous year. Sea World veterinarian David Kenney had justified the removal of "Gigi" from her mother in Baja California's lagoons as a unique opportunity for physiological research that would have been difficult or impossible with a wild, unconfined whale and would ultimately aid the conservation of her species. Because of this unprecedented research opportunity, some marine mammalogists, including Hubbs, were enthusiastic supporters of the capture; others, including Evans, were initially skeptical about its scientific and conservation value. Donald Wilkie, the director of the Scripps Institution's public aquarium and museum, had told Kenney immediately after Gigi's capture in March 1971 that he did not "think that it is necessary to justify the expedition on the basis of 'saving the species'"; the scientific and educational benefits of the capture alone were sufficient justification. Dubious conservation claims would ultimately undermine Sea World's case, Wilkie suggested, adding that he was "sure you will agree that the last thing Sea World wishes is to have the public question the credibility of its scientific staff."[12]

Before her capture Sea World had contemplated keeping Gigi in captivity indefinitely, but by the beginning of 1972 she had grown to over 26 feet long and over 12,000 pounds. Faced with the increasingly daunting cost of satisfying her appetite and providing a sufficiently large enclosure, as well as ongoing criticism from animal protectionists, Sea World applied to the Fish and Wildlife Service for a permit to release Gigi in March, when migrant gray whales would be passing the southern California coast on their way to northern feeding grounds. The permit was issued at the end of January 1972 on the condition that the release would be a joint effort with the Naval Undersea Center under Evans's direction. For Gigi's safety and to avoid adverse publicity should anything go wrong with the transportation and release, Charles Lawrence of the Fish and Wildlife Service strongly suggested to Sea World "that there be *no* advance publicity of the plans for (or time of) release of the animal."[13]

Radio tagging was central to Evans's plan for the release. In a memo outlining the plan to the U.S. Navy and Sea World staff that would carry it out, Evans noted that no one—not even those who had brought a much smaller Gigi into captivity a year earlier—had experience transporting a live whale of Gigi's size. Even if the transfer was successful, her chances of survival were uncertain. Taken from the wild as an infant, she lacked experience with migration, feeding, and social interaction with other members of her species and was almost certainly in poor physical condition compared to wild whales of her age. Radio tagging would, Evans explained, serve two purposes: monitoring Gigi's success in adapting to the wild and generating data that might be useful for gray whale conservation. On the basis of their efforts to radio-tag small cetaceans, Evans and his colleagues had concluded that surgical attachments caused "the least amount of stress to the animal." It was vital that the release be conducted with an eye toward the collection of scientific data, Evans added, since the "only justifiable reason for procuring and studying an endangered species is to provide knowledge which will aid the scientific community in arriving at sound procedures for wise management, protection, and propagation of the populations of these species still surviving."[14]

Under Evans's guidance, Gigi's release in March took place without incident. Despite some damage to the antenna of her bulky radio tag caused either when Gigi rubbed against the sea floor or swam through dense coastal kelp beds, the tag allowed her to be tracked up the California coast until May, when her signal was lost near Monterey. Although the tracking period was shorter than the six months that Evans had hoped for, it was by far the longest that any large whale had been tracked. Gigi's radio tag also made it possible to reassure the public and the press that she had not been harmed by her year in captivity. In contradiction to his formal statements at the time, which emphasized the tag's scientific purpose, Evans later recalled that the main reason Gigi was radio-tracked was to "assure the media that she was safe on her way north." Gigi's capture and release could easily have been seen as a failed commercial experiment by Sea World on whether a gray whale could attract enough visitors to pay for her own maintenance; Evans's participation helped transform it into a story about "a baby whale borrowed for science and returned to the sea."[15]

In April 1972, while Gigi was making her way northward under the release team's watchful eye, the House passed a version of Anderson's less restrictive marine mammal protection bill. The editorial board of the *New York Times* criticized the House for catering to the economic interests of the tuna-fishing industry rather than fulfilling the desire of "the public" for "a bill to stop the killing of ocean mammals," but the advocates of "rational management" eventually won out over the advocates of "strict protection" in the Senate as well. On October 21, 1972, President Nixon signed the bill into law as the Marine Mammal Protection Act of 1972. While the act established a blanket moratorium on the take of marine mammals, where "take" was defined as harassment, injury, or killing, it also established a system of permits and waivers that would allow such take to continue for a variety of reasons, including scientific research. A three-member Marine Mammal Commission and a nine-member Committee of Scientific Advisors were established to advise the National Marine Fisheries Service and the Fish and Wildlife Service on the issuance of permits and waivers. NMFS, located in the Department of Commerce, was responsible for implementing the law as it applied to whales, dolphins, seals, sea lions, and fur seals, while the Fish and Wildlife Service, located in the Department of Interior, was responsible for sea otters, manatees (and, later, the closely related dugongs), polar bears, and walruses. The passage of the Marine Mammal Protection Act was only the beginning of a long-running struggle over its implementation. Early in 1973, a few months after the act had technically gone into effect but before any of the new regulations and agencies it required were in place, Hubbs told conservationist Scott McVay that although he welcomed new legal protections for marine mammals, he hoped that the MMC and the relevant agencies would establish "reasonable regulations so that the occasional taking of moderate numbers for research, education, and exhibit can continue."[16]

Recognizing that the MMC would not be immediately operational and thus would not be able to advise NMFS and the Fish and Wildlife Service on the issuance of permits as required under the law, the drafters of the MMPA had provided for temporary exemptions on the basis of "hardship." In January 1973 Norris received one of these hardship exemptions in order to attach what the notice of his exemption in the *Federal Register* described as "telemetric data and tracking packages" to gray whales in one of Baja California's lagoons. In February 1973, he temporarily restrained and tagged several young whales using radio-tracking equipment designed by an engineer at the Franklin Institute Research Laboratories in Philadelphia. Although the "harnesses and capture worked to perfection," Norris later told Hubbs, the results fell far short of those Evans had achieved with Gigi; Norris had only been able to track the whale he had dubbed "Carl" in honor of his mentor for five hours before the tag broke loose. Though Norris had little trouble obtaining his exemption, many other marine mammalogists were frustrated to find that their work had suddenly been rendered illegal. At a congressional oversight hearing for the MMPA held in Anchorage late in the summer of 1973, marine mammalogist George Harry complained that, with only a few exceptions, "funds are presently being impounded and research has stopped." The director of NMFS later admitted that "practically all scientific research involving the taking of marine mammals" had "faced abrupt termination" when the MMPA went into effect, and that NMFS had granted only fourteen hardship exemptions for scientific research between the time when the law went into effect in December 1972 and the time when the regular permit review process became operational in the fall of 1973.[17]

Although Victor Scheffer, John H. Ryther, and Starker Leopold had been appointed commissioners of the Marine Mammal Commission in May 1973, with Scheffer serving as chairman, no money was appropriated for MMC staff or offices until November 1973. The operation of the MMC was further compromised during the first year of the MMPA by Scheffer's struggles with the House oversight committee to define the commission's role. Scheffer's career had begun at the Bureau of Biological Survey in the late 1930s, when he had accompanied Olaus Murie on a research expedition to the Aleutian Islands, and had continued with positions as a marine mammalogist with the Fish and Wildlife Service and the Bureau of Fisheries, the predecessor of NMFS. Compared to most of his colleagues, Scheffer was more sympathetic to protectionist and humane attitudes toward marine mammals and more critical of scientists' ties to theme parks, the fishing and whaling industries, and the military. When Scheffer's name appeared on a list of potential candidates for the MMC late in 1972, one former graduate student of Hubbs's admitted that he was well qualified but also disparaged him as a "naive idealist."[18]

Scheffer's opposition to the use of marine mammals for warfare brought him into conflict with the significant fraction of the members of the MMC's Committee of Scientific Advisors whose careers had been supported almost entirely by the Office of Naval Research. In November 1973, after he had criticized the U.S. Navy's Marine Mammal Program in an article in *Smithsonian* magazine, Ray sent him an irate, heavily punctuated letter defending marine mammalogists' collaboration with the military: "It is a plain fact that the Navy has done or has supported more basic ecological research on marine mammals and has developed more tools for the study of them than the NMFS and the National Science Foundation combined!!" A few months later, when Scheffer appeared before the House oversight committee, he was forced to admit that he had a "personal, private conviction that at every stage, and continuously, we should try to whittle down the growing power of the military in areas where it has no business to be," though he also claimed he would not allow that conviction to bias his work for the MMC. Such statements endeared Scheffer to antiwar and animal welfare activists, but they alienated him from many of the marine mammalogists and policy makers with whom he had to work on a daily basis.[19]

More broadly, Scheffer believed that "the public" deserved a voice in marine mammal policy at least as strong as that of scientists, industry, and the military, and that one of the purposes of the MMC was to amplify that voice. In late September 1973, he explained to a colleague who had recently complained to the Smithsonian's Dillon Ripley that neither Leopold nor Ryther were marine mammal specialists that he did not share his concerns. On the contrary, he thought marine mammalogists were overrepresented on the MMC's scientific advisory committee and hoped soon "to see women represented on the Committee, as well as persons having a broad concern for the welfare of marine mammals, though lacking conventional training in marine mammalogy."[20]

In the meantime, Scheffer attempted to limit the role of the advisory committee in the MMC's decisions. Even after John Dingell, the chairman of the House Subcommittee on Fisheries, assured Scheffer in October 1973 that the MMPA directed the MMC to consult its Committee of Scientific Advisors "in every case where the facts available to the Commission are in dispute or unknown," Scheffer continued to tell the advisory committee that, according to his own reading of the law, they would only be consulted "on matters having scientific implications," such as permit applications for scientific research. When Schevill, a concerned member of the advisory committee, wrote to the subcommittee for clarification, Dingell reaffirmed his instructions to Scheffer in even stronger terms: because the MMC "was never intended to consist of marine mammal experts," consultation with its advisory committee was essential on all matters. Otherwise, he told Scheffer, its decisions "might well and correctly be viewed with suspicion by the public."[21]

Scheffer's attempt to position the MMC as a voice for "the public" continued until an oversight hearing held by the House Subcommittee on Fisheries in January, where committee counsel Frank M. Potter Jr. definitively rejected his interpretation of the law. A few months earlier, Potter had written Scheffer to explain that "the Commission is not a spokesman for the public and should avoid getting trapped in that role. If you are a spokesman for anyone, it is for the animals." At the oversight hearings, Potter asked Scheffer if he continued to believe that one of the responsibilities of the MMC was "to serve as an agent of mediation between the public and public servants when disputes over conservation of marine mammals arise," as he had written in a paper he planned to deliver at a conference the following month. When Scheffer said he did, Potter said that whatever his personal convictions might be, he was misconceiving the role of the MMC. Far from being a spokesman for the public or a middleman between the public and government agencies such as NMFS and FWS, he told Scheffer, "You are, in fact, the disinterested expert."[22]

Though Scheffer was given little opportunity to discuss substantive issues of marine mammal science or policy, he did raise a number of questions he believed deserved the focused attention of the MMC in his opening statement. Among them was a question of concern to both scientists and animal welfare activists: "At what point does research, such as tagging and counting entail harassment?" In advance of the January 1974 oversight hearings, Christine Stevens of the Society for Animal Protective Legislation and the Animal Welfare Institute had written to the director of NMFS to question several permits to tag marine mammals that had recently been issued and to suggest the need for "thorough and continuing scrutiny" of the effect of tagging on marine mammals. She reiterated her concerns at the hearings, telling the committee that the use of Discovery marks was especially unfortunate, since it amounted to a tacit approval of the whaling industry, but that even tags that did not depend on whalers for recovery were problematic. Marks that penetrated blubber and muscle were clearly painful to individual whales, she argued, though "how painful is extremely difficult to know." When Potter asked Stevens whether she thought that "research per se is bad," she replied that she merely wanted to ensure that research interfered with the animals as little as possible. Potter agreed that "research should be very heavily weighted toward nonconsumptive research, although that may not always be possible, and it should be very carefully designed so that if you are going to have to go in and disturb animals which are not present in abundance, you do the absolute minimum necessary," but he questioned the need for the stricter scrutiny of scientists that Stevens had called for. The marine mammalogists he had spoken with about marking and tagging, he told Stevens, claimed that it was in their own interest to change the behavior of the animals they studied as little as possible. "To that extent, I think we all are in agreement," Stevens replied, "but scientists often have to be reminded, in my experience."[23]

Humaniacs and Extremists and Intimidators

On February 4, 1974, notice of an application from Sea World to NMFS for a permit to capture four killer whales in the Puget Sound area appeared in the *Federal Register*. That spring, while NMFS and the MMC reviewed the application, University of Washington wildlife biologist Albert W. Erickson visited the theme park in San Diego to speak with its leadership about the possibility of conducting a radio-tagging study in the course of the planned collection operation. Sea World had been considering such a collaboration since at least the previous fall, when Evans had sent copies of research proposals he and Douglas G. Chapman, a whale researcher also at the University of Washington, had drawn up for studying marine mammal populations to Sea World's general manager, Frank Powell. Although Evans had included the use of freeze brands, visual tags, and radio tags in his own proposal for censusing southern California's dolphins, he encouraged Powell to begin research on Puget Sound's killer whales with a "good well designed aerial and ship census." Tagging could wait until "maybe phase II or III."[24]

In contrast to Evans, Erickson was eager to begin tagging immediately. Much of Erickson's career as a wildlife biologist had in fact been devoted to the development of techniques for handling and tagging dangerous animals. In the late 1950s, he had pioneered chemical immobilization and live trapping of black bears in Michigan, inspiring the Craigheads, Dean, and others to use similar techniques on grizzlies, and in the late 1960s he had worked closely with Siniff to develop techniques for handling Antarctic seals. (Siniff later described Erickson as a "madman" who had no fear of working under dangerous or extreme conditions.) In a letter written after his visit to Sea World, Erickson told Powell that he was optimistic that the MMC would fund a radio-tagging study of killer whales, especially since Chapman, a member of the MMC's scientific advisory committee, was helping him design the study. In exchange for Sea World's help, Erickson offered to assist with a new exhibit on polar animals. His experience in capturing seals, he told Powell, would be helpful "in gaining clearance for making the collections" from the MMC, and he believed the permit application would be further "enhanced" by proposals from outside scientists such as himself to study the animals in captivity. Finally, his credibility and authority as a scientist would be helpful in fending off the criticism from animal protectionists that was likely to be directed toward the new exhibit: "Considering the tenor of the day, a prime concern of any polar exhibit utilizing polar mammals would be in defending any proposed action and in this function I trustfully might also be able to assist Sea World."[25]

In May 1974, about three months after the initial notice of Sea World's permit application had been published in the *Federal Register*, NMFS issued the permit, albeit with significant restrictions on where and how the capture of the four orcas could be performed. Erickson's proposal to the MMC for funding to radio-tag killer whales was also successful. In the permit application that he submitted to NMFS in the spring of 1975, Erickson proposed to "take" ten killer whales in collaboration with Sea World's collecting operation. The focus of the study, Erickson told NMFS and the MMC, was on techniques rather than on scientific questions; its aim was to develop new visual marking techniques and a "reliable radio-tag." Two of the orcas that Sea World captured temporarily but decided not to bring into permanent captivity would be kept for up to two months while Erickson tested various marking methods, after which they would be released back into Puget Sound. If the tagging and tracking methods proved successful, they would then be used to study as many as eight other whales. NMFS awarded Erickson's scientific research permit in June 1975, less than three months after he applied for it and well before the first opportunity to radio-tag killer whales that met the requirements of Sea World's collection permit presented itself.[26]

Though Erickson's research was not significantly delayed by the new permitting requirements of the MMPA, many scientists and public display institutions such as Sea World were becoming increasingly frustrated at having to wait for months to find out whether their permit applications had been approved. Even the director of NMFS, Robert Schoning, admitted that the permit process had been unacceptably slow when he appeared before the House oversight committee in the fall of 1975. However, the average processing time for the twenty-five permits issued since the beginning of 1975 had fallen to three months, Schoning told the committee, and was expected to fall further as NMFS refined its review process. Marine mammalogists frustrated by the permit process found an ally in Sea World, whose vice president for research and veterinary husbandry, Lanny Cornell, testified on behalf of the American Association of Zoological Parks and Aquariums at the same hearing where Schoning admitted that NMFS's performance still needed improvement. Without mentioning Erickson's successful permit application earlier that year, Cornell told the committee that the MMPA's permit requirements had "stifled totally" scientists' ability to conduct "simple spontaneous field research," such as freeze-branding, the extraction of teeth, or the attachment of radio tags to animals incidentally encountered in the process of doing research or collecting for public display. Unless Congress took prompt action, Cornell told the committee, the progress of marine mammalogy would be halted by "hysterical dictating by a vocal and emotional few who could by their actions cause the knowledge of all animals with whom we share this earth to return to the dark ages."[27]

The use of such hyperbolic language to describe the threat of animal protectionists was not unusual for the representatives of the public display industry speaking before Congress or to the press. A representative of the Zoological
Action Committee who also testified at the fall 1975 oversight hearing accused
NMFS of capitulating to "intimidation" from "humaniacs" such as Alice Herrington of the New York–based organization Friends of Animals, who had testified in several previous hearings on marine mammals. The committee took such
rhetoric with a grain of salt. In response to the testimony attacking Herrington,
the chairman of the committee noted that it would "try to relate to the industry
and the humaniacs and extremists and intimidators" in the hope that it could
"reach with due deliberation some reasonable result for all parties concerned."[28]

The "vocal and emotional few" that Cornell had identified as the enemies
of progress were represented at the hearings by Jane Risk of the Animal Protection Institute, who accused NMFS of issuing permits for studies that were
"inhumane and detrimental to the health and well being and behavior patterns
of the animals involved." One of the most objectionable permit applications, she
told the committee, was a proposal to radio-tag bowhead whales in the Bering
Sea that had been submitted earlier that year by Ray and one of his colleagues
at Johns Hopkins University. In May, the Monitor Consortium, an alliance of
animal protection and conservation groups, had asked Schoning to postpone
a decision on the permit until a public hearing had been held on the possibility of injury to the whales. According to the permit application, the projectile
had been designed to lodge in the skin and blubber rather than the muscle, but
the researchers had failed to discuss the implications of the antenna protruding
through the skin, including "possible vibration and consequent movement of
connecting hardware in the animal's skin during swimming, turning, stopping,
and starting causing ulcers in the skin." Like Stevens before her, Risk was careful
to reassure the oversight committee that although the Animal Protection Institute opposed some of the permits that NMFS had issued, it was "for research,"
not against it, and would support tagging methods that had been tested and
shown to be benign. She was not a "humaniac," she told the committee, and did
"not think name calling will serve anything."[29]

Sea World's first opportunity to capture killer whales under the conditions
of its permit, which limited the geographical regions of Puget Sound where
the capture could take place and required the presence of an NMFS observer,
came on March 7, 1976, at the Budd Inlet near Olympia, the capital of Washington State. On what was later described as a "good sailing day," the capture

team followed a pod of orcas into the inlet and then drove them close to shore with boats, nets, and small explosives. The operation was witnessed by several hundred people on shore and in boats, including Ralph Munro, a high-ranking member of the staff of the governor of Washington, and some of the attendees of the world's first conference on killer whales, which was then under way in Olympia. It was immediately met by a wave of protest that included several of Washington's most prominent political figures. As the *Marine Mammal News* noted, attitudes toward the capture of killer whales in the Pacific Northwest had changed dramatically since the first live capture in Puget Sound in 1965, when "the leader of the collecting expedition was welcomed into Seattle as a hero and given the key to the city." Immediately after the Budd Inlet capture, at Munro's urging, Washington State's attorney general filed suit against Sea World and NMFS in federal court, claiming that the state had not been properly consulted before the issuance of the permit. On March 11, 1976, a temporary injunction was imposed on Sea World not to transfer the whales out of the Budd Inlet until the suit had been decided.[30]

Recognizing that Washington State's authority over killer whales in its coastal waters had been severely limited by the MMPA, the state's congressional delegation also sought to amend the act to prevent further captures. On the day the injunction was issued, Warren Magnuson introduced a bill into the U.S. Senate to amend the MMPA to prohibit the taking of killer whales for the purpose of public display. Although the bill made an exception for scientific research, the version that was eventually passed by the Senate prohibited any research that involved removing an orca from the water or endangering its health or well-being. Either of these qualifications would have prevented Erickson's study from being authorized. In his defense of the bill to his colleagues in Congress, Magnuson noted that the "residents of Puget Sound have developed a great attachment and concern for killer whales," and that the capture operation in the Budd Inlet had produced "outrage" that had "continued unabated ever since." While most of the outrage was directed against capturing what Magnuson called "one of the most marvelous animals in existence," he noted that some people also objected to Erickson's proposed tagging study. His own feeling was that some research was necessary but that it was "absolutely inhuman treatment . . . when they are put in a pen." When Senator Strom Thurmond questioned the need to single out killer whales for special protections without holding hearings to consider the matter carefully, Magnuson explained that the situation was "a kind of emergency." Two of the killer whales had escaped from their temporary sea pen, possibly injuring themselves in the process, while others were "now up at the University of Washington. They will be releasing them and letting them go out to sea after they put some kind of radio in their fins. This is the way they treat them, but they cannot live that way." Research had shown, Magnuson added, that orcas suffered when their close-knit family groups were torn apart. When young orcas were removed from their mothers, "it makes you sick to your stomach to listen to them cry. They cry like babies, or vice versa."[31]

Despite Thurmond's reservations, Magnuson's bill was passed by the Senate with little debate. On the same day, Sea World and Washington State reached an out-of-court settlement; the state's suit was dropped in exchange for Sea World's releasing the whales it had captured and promising never to attempt to capture killer whales in Washington again, even if it was legally allowed to do so under the MMPA. Washington's state legislature also passed a law banning the capture of killer whales in its waters, following the lead of British Columbia, which had passed a similar law in the fall of 1975 after the controversial capture of a pod of orcas by Sealand of the Pacific, an aquarium located on Vancouver Island. On the federal level, Bob Bonker, a representative from Washington State, introduced the equivalent of Magnuson's killer whale bill to the House, noting at a hearing on the bill held in early May that although much remained unknown about killer whales, he could "personally testify to one incontrovertible killer whale fact—when people see a modern killer whale hunt, they do not like it. The display of public outrage is immediate and widespread and the demands for action on the behalf of the whales are persistent." Milton Kaufman of the Monitor Consortium testified at the hearing that outrage against breaking up orca family groups was widely shared; it went well beyond "the frequently derogated protectionists or those 'emotional humaniacs.' "[32]

By the time the hearing was held, however, the emergency had passed, and Sea World's allies in the scientific community had rallied to support the legitimacy of capturing cetaceans from the wild. In late March, Hubbs had telegraphed California's congressional representatives as "a knowledgeable authority on marine mammals, a strong and effective conservationist for decades, and Board Member of Sea World" to protest Magnuson's "ill-advised" bill, which "if passed would be a critical blow to Sea World and other authentic marine exhibits, also to Sea World's new Research Institute and to an important industry and tourist attraction in San Diego, Ohio, and Florida." At the hearing in May, Milton Shedd, chairman of Sea World's board of directors, stressed the scientific benefits of the Budd Inlet capture operation, notably Erickson's project "to tag these whales with radio transmitting devices in order to further compile much needed data relative to the population dynamics and migration patterns of killer whales." Evans and one of his colleagues in the U.S. Navy's Marine Mammal Program, Sam Ridgway, also testified against Bonker's bill. When asked about their relationship to Sea World, they both denied that they had "any connection, professionally or otherwise, with any commercial whale display organizations," a statement that elided the extensive if informal interdependence between the military, academia, government agencies, and commercial oceanariums that characterized almost all cetacean research at the time, but especially that conducted in San Diego. Evans told the committee that the "controlled experiments" possible only with captive animals were invaluable and that techniques for capturing and keeping killer whales were continuously improving. Seemingly redundant in light of the settlement between Sea World and Washington State and assurances from the MMC that it would advise NMFS against issuing further permits to capture Puget Sound orcas until the population had been thoroughly studied, the Bonker bill stalled in committee after the hearings.[33]

Despite the controversy, Erickson was able to radio-tag and release two of the orcas captured in the Budd Inlet, a male and a female who had first been transferred to the Seattle Marine Aquarium and then to a pen near Kanaka Bay on San Juan Island. Through Siniff, Erickson was connected to the Cedar Creek Bioelectronics Laboratory, which designed a radio tag for the orcas using a plaster cast of the dorsal fin of Sandy, a captive killer whale at Sea World. Erickson had hoped to track the killer whales for as long as a year, but after ten days of intensive tracking he lost the signal because of radio interference, and after that the whales were tracked only intermittently. In his report to the MMC, Erickson suggested that the killer whale would "appear to be a species of choice in the development of satellite telemetry for cetaceans" because of its relatively large size compared to other species of small cetaceans, but neither he nor anyone else pursued the possibility; Erickson did not even attempt to tag the remaining eight killer whales allowed by his permit. The July 1976 issue of the *Marine Mammal News* reported that despite "fair success" with radio tracking, Erickson was exploring alternatives, such as tracking the whales using their vocalizations. With the technologies available at the time, capturing killer whales so that they could be radio-tagged was an expensive and dangerous endeavor that scientists could not easily accomplish without the help of Sea World and other commercial collectors, which were now excluded from Puget Sound.[34]

In response to protest against the Budd Inlet capture and criticisms of its performing animal exhibits, Sea World intensified its efforts to portray itself in the media as a site for science and conservation in the mid-1970s. One of the most visible components of this effort was the expansion and rebranding of the Mission Bay Research Foundation, which had been supporting marine mammal research in San Diego on a small scale since the early 1960s, as the Sea World Research Institute. Evans later recalled that the creation of the institute was a direct response to the passage of the MMPA, which required public display institutions that held marine mammals to develop research and education programs. In 1975, Richard B. Lippin, a representative of a public relations company hired by Sea World to promote the new institute, told Powell that the "overall objective of our public relations program will be to establish Sea World as an important contributor to man's understanding of the marine environment through the many projects of the non-profit Sea World Research Institute." Among the target audiences of the program, Lippin explained, would be "influential individuals in government, whose recognition of Sea World as a unique research facility should be of considerable value in legislative matters."[35]

In 1977, after having spent most of the early 1970s completing his doctoral thesis on small cetaceans and helping the U.S. Navy's Marine Mammal Program prepare its first permit application under the MMPA, Evans became the first executive director and senior scientist of the newly renamed Hubbs–Sea World Research Institute. Researchers at HSWRI were funded by a variety of sponsors including, during Evans's first two years, the National Science Foundation, the Office of Naval Research, the MMC, and the California Department of Fish and Game. ONR, for example, awarded Evans a contract for the development of radio tags for small cetaceans to begin in the spring of 1977. Meanwhile, Sea World provided HSWRI with a significant portion of its funding, a plot on the land it leased from San Diego, veterinary consultation, and three of its ten board members. In exchange it gained the right to claim HSWRI's scientific accomplishments as its own. As scholar Susan Davis points out, Sea World and HSWRI were legally independent entities, but they were tightly connected in practice and became virtually indistinguishable in Sea World's public relations materials. The title of a promotional pamphlet produced in 1977, which described the work of the new research institute and included a page on Gigi's capture and release, was typical of Sea World's ongoing attempt to transcend its commercial roots: *Three Worlds of Sea World: Entertainment, Education, Research.*[36]

Although Erickson had told the press that he was no longer pursuing the radio tracking of killer whales, other biologists continued to hope that the technique would be used again. In 1977 Chapman drafted a report for the MMC's Committee of Scientific Advisors that argued for the return of a "live-capture fishery" for Puget Sound's orcas, albeit under significant restrictions. In the spring of 1977, Siniff urged Chapman to emphasize the necessity of radio tagging and other kinds of artificial marking. Siniff was skeptical of claims that the method of identifying killer whales on the basis of photographs of dorsal fins and distinctive patterns of coloration that had been developed by Canadian researcher Michael Bigg and others was sufficient for understanding killer whale demographics or that radio tagging was as expensive or ineffective as some critics had claimed. Erickson's study, he told Chapman, had proven otherwise. The draft that Chapman later sent to Siniff suggested that it was "ironic" that the success of Sea World and other theme parks and aquariums in developing public appreciation for killer whales "had produced such a strong counter-reaction against its own capture operations" and that this concern about capture had overshadowed more pressing environmental threats to the orca population. Following Siniff's recommendation, Chapman argued that some important questions about the Puget Sound population could only be answered through radio tracking, which would probably only be feasible if conducted as part of a live-capture operation. After reading the draft in the fall of 1977, Siniff told Chapman he wished it had made "a stronger pitch for marking, but in view of the delicate situation, I suspect we should leave it like it is currently written. It seems quite certain that research involving radio telemetry and marking will only be accomplished in the presence of a live-capture fishery. Therefore, it seems unlikely that this will ever be done in Puget Sound and I guess I think that is a shame." When Robert Eisenbud, general counsel for the MMC, reviewed the draft the following year, he warned Chapman that its recommendations might launch another round of the killer whale controversy, which was now "dormant."[37]

Despite the skepticism of Siniff and others, photo-identification studies quickly became the primary method of studying the killer whales of Puget Sound and the area around Vancouver Island, including the two orcas radio-tagged by Erickson, who were easily identified by the scars that the tags had left when the bolts with which they were attached tore through the orcas' dorsal fins. The photo-identification studies were coordinated by Bigg and Kenneth C. Balcomb III of the Moclips Cetological Society in Friday Harbor, Washington, with contributions from NMFS's National Marine Mammal Laboratory in Seattle, but they involved a broad network of observers. Anyone with a camera could, in theory, send a photograph of a killer whale to Balcomb or Bigg, who could then match the photo against their databases of dorsal fins. Whereas Erickson's radio-tagging study had clashed with widespread interest in killer whales, photo-identification research thrived on it. According to Robin Baird, a killer whale researcher of the generation following Bigg's, Bigg had radio-tagged a killer whale near Vancouver Island in 1973, three years before Erickson's attempt, but had only been able to track it for eight hours before losing the signal. Baird argues that this failed experiment was "an important event in the subsequent tagging studies of killer whales," since Bigg dominated killer whale research for the next two decades. His conviction that the technique was ineffective was one of the main reasons, Baird suggests, that "subsequent radio tracking studies with killer whales have lagged behind applications to other species of marine mammals."[38]

Let Whales Alone

In the late 1970s, after orcas in the coastal waters of Washington State and British Columbia had been placed off limits by state laws and public protest, Sea World and other theme parks and aquariums in search of killer whales shifted their attention to Iceland, where killer whales were abundant and public opinion had not turned against capture. More than fifty-five orcas were taken into captivity in Icelandic waters between 1976 and 1989, many of which were kept in Iceland for several years before being exported to the United States, Canada, and other countries. Even when, because of a loophole in British Columbia's law against the capture of killer whales, Sealand of the Pacific was issued a permit in 1982 to capture two local orcas as replacements for one captive orca that had died and a second that the aquarium planned to release, Icelandic orcas proved easier to acquire. Upon learning of the permit, Greenpeace activists in Vancouver and Seattle promised to block the planned capture operation using direct action if necessary. Originally founded in Vancouver, the environmental organization took a close interest in killer whale protection; when the design for a new Greenpeace logo debuted on the masthead of the *Greenpeace Chronicles* newsletter in the fall of 1975, it was intertwined with the flukes of a killer whale. Before the captive orca could be released, enabling Sealand to capture his replacement from the wild, he died, and Sealand's capture permit was revoked. Soon afterward the theme park imported three orcas from Iceland, one of whom died soon after arrival.[39]

Though radio tagging had been tarnished in the case of killer whales by its association with the Budd Inlet capture and seemingly rendered unnecessary by the photo-identification work of Bigg and his colleagues, the scientific value and ethics of using the technique to study cetaceans remained the subject of intense debate. One of the most vocal advocates of "benign" techniques of research was Payne, who had collaborated with conservationist Scott McVay to popularize recordings of the songs of humpback whales in the 1970s. In 1983, Payne published a collection of papers describing studies he and his students and collaborators had recently conducted using "passive observation techniques," none of which involved "killing, capturing, confining, or even touching a whale." While there were many "intrusive" techniques that were worthwhile and in some cases might be the only way to answer questions important for management, Payne argued in his introduction to the volume, "benign research techniques" were far more informative than most scientists believed. Still, although none of the studies in the book involved handling whales, and Payne and his coauthors on a study of the photo-identification of right whales argued strenuously that artificial marks were less useful than most scientists thought, they were careful to note that they did not oppose radiotelemetry or other kinds of "active tags," such as the acoustic tags with which Payne had experimented in the 1960s. Indeed, they wrote, there "are certainly many things that can be learned only through radio tags." Radio tracking thus occupied an ambiguous position in discussions of "benign" research on cetaceans. Bolting a radio tag to a temporarily restrained orca might seem harmful compared with taking a photograph or recording vocalizations with a hydrophone, but it could easily seem benign when compared to killing or permanent captivity.[40]

The exact meaning of the term "benign" was one of the recurring subjects of discussion at a conference at the New England Aquarium in Boston in the summer of 1983, formally titled the Global Conference on the Non-Consumptive Utilization of Cetacean Resources but more widely known as "Whales Alive." In his paper there, Payne welcomed the increasing use of "benign research techniques for field studies of free-ranging cetaceans" over the past decade but also argued that "benign research" should be understood to encompass both "non-consumptive" and "low-consumptive" methods. The latter category included studies involving capture or serious harassment, as well as studies that relied on the "assistance" of whales to collect environmental data, including the attachment of satellite tags to migratory whales as "a relatively cheap means of collecting basic oceanographic data from remote ocean areas." Payne's understanding of "benign" was shared by a number of other participants in Whales Alive. In his summary of the discussion, Robbins Barstow, the organizer of the conference, defined benign research as "research that does not depend on the human-caused death of wild animals nor involve significant stress or injury to them," a definition that he believed allowed for the use of "aerial photography, radio tracking, visual camera scanning, listening to whale sounds, and satellite telemetry."[41]

Not everyone at Whales Alive agreed that "non-consumptive" or "low-consumptive" use of whales was justified, however, even for the purposes of science or conservation. The philosophers Dale Jamieson and Tom Regan presented a paper titled "Whales Are Not Cetacean Resources: An Animal Rights View," in which they challenged the conference's premise that "non-consumptive" uses of cetaceans, such as whale-watching, should be promoted over "consumptive" uses, such as whaling. The problem was not how consumptive the use was, they argued, but the fact that the relationship between humans and cetaceans was framed as "use" in the first place. Jamieson and Regan argued that treating whales as objects or as means to human ends deserved "moral skepticism," even in an activity as "apparently harmless . . . as aerial viewing for the purposes of population estimates," let alone a technique as invasive as radio tracking: "Just as whales are not here for us to kill for our purposes, so they are not here for us 'to study,' or 'to watch,' or 'to play with.'" Even research that would help protect whales as a class was to be regarded with skepticism, since we already knew enough about them to know that individual whales should not be treated as objects of research. Instead our ethical responsibility was simple: "to *let whales alone.*"[42]

In the spring of 1983, after having focused its orca acquisition efforts on Iceland since the Budd Inlet fiasco of 1976, Sea World applied to NMFS for a permit to "take" one hundred killer whales off the coast of California or Alaska over a five-year period. Up to ten of the whales would be brought to Sea World's three parks, where they would be trained to perform and contribute to a nascent captive breeding program, while the other ninety would be kept in temporary captivity for as long as three weeks, to be "sampled, marked, and/or tagged and released" for scientific purposes. That summer, Greenpeace began urging its members to write letters to NMFS in opposition to the proposal. A particular target of Greenpeace's attacks was Sea World's attempt to frame the operation as primarily a scientific research project. Greenpeace both dismissed the scientific value of the proposed studies and questioned, as Regan and Jamieson did at that summer's Whales Alive conference, whether humans had any right to treat killer whales as "specimens," regardless of the potential gains to science. Sea World's emphasis on scientific research was, Greenpeace argued, a transparent attempt to avoid the "messy protests" aimed at Sealand's "blatantly profit-oriented efforts" during the previous year. A truly research-oriented proposal would have relied on the "benign" methods of field observation and photo-identification used by most killer whale researchers or on newly proposed methods for obtaining tissue samples for genetic analysis using biopsy darts, neither of which required capture. To bolster its case, Greenpeace told its members that it had the support of no less an authority than Scheffer, the MMC's former chairman, who had told the press that he opposed the issuance of the permit because the "killer whale is an outstanding symbol of freedom and wildness. A killer whale in a pool is no longer a killer whale."[43]

In July 1983, under pressure from environmental activists and from Washington State's congressional delegation, NMFS announced that it would hold a public hearing about Sea World's permit application in Seattle the following month. It also announced that it would continue accepting public comments until ten days after the hearing, which would be more than five months after the notice of application had first been published in the *Federal Register*. In the days before the hearing, Sea World attempted to undermine Greenpeace's accusations of unbridled greed. Cornell told the press that Sea World's scientific and financial motives were entirely compatible; the study would "help the scientific community learn a lot about an animal of which we know very little, while allowing us to present animals that the public wants to see." While Greenpeace's report on the hearing claimed that "the vast majority of participants spoke against the project, accusing Sea World of couching its real motive, greed, behind grandiose talk of scientific mission," a number of supporters of the proposed capture and research program also testified. Among them was Erickson, whose testimony about the scientific value of the proposal was cited by the editorial board of the *Los Angeles Times* when it endorsed Sea World's plans in September.[44]

Despite the opposition of animal protection and environmental groups and many killer whale researchers, NMFS issued the permit to Sea World at the beginning of November. An agency representative later told the House subcommittee with oversight authority for the MMPA that NMFS had decided that "the permit would be consistent with the purposes and policies of the act" only after reviewing "thousands of comments both in favor of and opposed to issuing the permit to Sea World." The knowledge gained from research on captive and wild animals would, he added, be "invaluable" for conserving killer whales in the wild, and the extensive conditions and restrictions included in the permit would "prevent any significant adverse impacts on the populations of killer whales involved." In addition to restricting capture activities to Alaskan waters, the permit placed limits on the number and location of animals to be captured each year and required annual review and reauthorization by NMFS, as well as an immediate cessation of capture activities in case of the death of a whale. It prohibited "extended temporary removal" of killer whales beyond a three-week period and denied permission for a variety of procedures that Sea World had planned to perform on temporary captives, including liver biopsy, stomach lavage, tooth extraction, hearing tests, and respiratory studies. Branding and radio tagging, in contrast, had been authorized as essential for compliance with the permit; they would allow Sea World operatives to determine whether a particular whale or pod had been captured previously. Finally, the NMFS representative told the House subcommittee, granting the permit was in accord with the MMPA's recognition of the important role of public display institutions such as Sea World in providing "an opportunity for millions of people to observe, enjoy, and learn about marine mammals each year."[45]

The extended public review of Sea World's proposal and the significant restrictions in the issued permit did not convince opponents of the legitimacy of the process. Soon after the permit was issued, Rod Chandler, a representative from Washington State, introduced a bill to the House to amend the MMPA to prohibit the capture or import of killer whales for public display. In his remarks on the bill, which was virtually identical to those sponsored by Magnuson and Bonker in 1976, Chandler echoed Scheffer in describing killer whales as "magnificent symbols of freedom, grace, and power. They are highly intelligent, sensitive, and social mammals which should be protected from unnecessary harassment, pain and death." He took care to note, however, that his bill would not prohibit scientific research that involved temporary capture or tagging. Representative Norm Dicks, one of the cosponsors of the bill, later told the relevant House subcommittee that he agreed with Chandler that "exploitation, even in the name of scientific research, is still exploitation, and we should not allow it," but he assured the committee that the bill "would not prohibit temporary detention or tagging of whales for study and research."[46]

In late November 1983, not long after Chandler's introduction of the bill, Norris defended the propriety of keeping of small cetaceans in captivity and the right of marine mammalogists to conduct research as they saw fit, including the use of invasive or lethal techniques when necessary, in a speech at a conference on marine mammals in Boston. As the outgoing president of the Society for Marine Mammalogy, Norris told his audience of about four hundred people, he was "worried about the public perception of science in relation to the popular conception of marine mammals." He and others had founded the society several years earlier, he explained, in response to the increasing influence of "uninformed voices" claiming to speak on behalf of marine mammal science. Those voices threatened marine mammalogy with their arguments for the "unacceptability of 'terminal' experiments; the suggestion that scientists are harming animals by tagging them; the movement to halt all animal shows; and the belief of some that the only contact humans should have with marine mammals is through the movies." Government regulation posed another worrisome threat. The passage of the MMPA had brought much-needed protection to marine mammals but only at the high price of driving the "prominent scientist who did much for marine mammal science" away from field research through burdensome red tape. The most effective response to these challenges, Norris argued, would be for members of the Society for Marine Mammalogy to continue to emphasize the value of "professionalism" in interactions with policy makers, the press, and the public.[47]

Over the next several years, Norris became one of the most vocal defenders of keeping small cetaceans in captivity and using invasive techniques to study them in the wild. In a speech titled "The Dolphin Connection," a draft of which he shared with Chapman, he admitted that there was "right and wrong on both sides" of the debate between scientists and animal protectionists, but his sympathies lay with scientists. "None of our work with wild dolphins," he wrote, "has more than trivial impact on them. And the chance to capture, tag, age and release animals, and to track them by radio is needed both for reasons of basic science and management." If animal protectionists had their way, he argued, not only would studies of captive animals come to an end, but "wildlife managers too, would lose most of their tools. Pretty clearly it would no longer be acceptable to mark animals, to capture them for age and sex determination, or to sample blood or tissue for genetic or hormonal analysis of populations. All these things have become important aids in determining the boundaries of wild populations, and hence the units that must be protected." Norris agreed that "each of these scientific endeavors brings up humanitarian questions," particularly, "how much manipulation should society allow in the name of discovery, or even of the protection of the animals themselves?" But he believed that the answer to this question was far more manipulation than protectionists were willing to agree to.[48]

Norris's claim to speak on behalf of all marine mammalogists did not go unchallenged, particularly after he criticized the Whales Alive conference as unscientific. Sidney Holt, a British marine mammalogist, prominent critic of the International Whaling Commission, and participant in Whales Alive, told one colleague that he believed the stridency of Norris's rhetoric was a sign of the narrowness of the group he actually represented. His disparagement of the Whales Alive conference, Holt wrote, reflected the professional allegiances and anxieties of a "very small group, all from USA and all connected with the commercial oceanaria and/or the US Navy" who were "upset *precisely because* the participation was so varied and it was an event out of their control." Moreover, Holt argued, Norris's portrayal of defenders of cetacean captivity as rational "truth-sayers" and their opponents as "emotional" and uninformed were transparently ad hominem attacks that belied his claims to superior rationality. "I happen to think I care about science at least as much as you do," he told Norris directly in the spring of 1985. "What I do believe is that the social and other costs must be taken into full account when particular research activities are considered, and that scientists as such have no god-given right to take decisions about what they will or will not do without consideration of wider interests and consequences."[49]

When the House Subcommittee on Fisheries held hearings on reauthorization of the MMPA in the spring of 1984, Chandler testified that his amendment prohibiting the capture of orcas was not intended to put Sea World out of business or to prohibit scientific research on killer whales, as some opponents had claimed. He would not have introduced it, he told the committee, "if the environment of orcas could be duplicated in captivity or if captive research was the only scientific means to study orcas," but it was evident that orcas suffered in captivity, living "drastically shortened" lives and becoming "lethargic and sometimes neurotic." Although Sea World staff clearly had "genuine concern" for the welfare of their animals, he added, "there is a much deeper question of right and wrong, kindness and exploitation, involved in this issue." As for the value of the scientific research that Sea World had proposed to carry out as part of its capture operation, there were now, Chandler claimed, "proven methods for studying killer whales in their natural environment which do not require capture," including "sighting networks, photographic identification, and acoustic monitoring."[50]

Despite the provision in Chandler's bill allowing killer whales to be captured for the purposes of scientific research, many of the supporters of the bill opposed hands-on research of any kind. At the MMPA reauthorization hearings, Chandler submitted for the record a letter from two of his constituents criticizing the Sea World permit's authorization of "invasive techniques" such as "radio tagging, spaghetti and streamer tagging and cryogenic marking techniques," which, they argued, had been rendered unnecessary by the effectiveness of photo-identification and hydrophonic recordings. Paul Spong, a British Columbia–based killer whale researcher and activist who had been closely involved with Greenpeace's campaign against commercial whaling and who had been present at the killer whale symposium in Olympia during Sea World's capture operation in the Budd Inlet, also testified against the use of invasive research methods. In 1969, after studying a killer whale in captivity at the Vancouver Aquarium, Spong had told the press that he hoped to use radiotelemetry to study killer whales in the wild, but he now argued that that technique was unnecessary. There was tremendous scientific value in "pristine" populations of orcas, he told the committee, which were already being effectively studied in British Columbia and Washington State with "benign, non-intrusive research techniques" such as photo-identification. Maxine McCloskey of the Whale Center in Oakland, California, similarly testified in favor of "benign methods" that "cause a minimum of disruption to the stable family pods of orcas." Far from being necessary for understanding orca populations, she argued, capture-based studies might actually interfere with long-term research.[51]

These claims about the risks of invasive research, the sufficiency of "benign" methods, and the need to provide killer whales with protections beyond those already provided by the MMPA were challenged by NMFS, the MMC, and members of the MMC's Committee of Scientific Advisors. Chapman, for example, pointed out in his testimony that the issued permit had incorporated a number of recommendations from the advisory committee to reduce the risks of "invasive procedures" to individual whales and to minimize the effects of the operation on the population as a whole. In fact, the only invasive procedures that had been allowed beyond capture itself were freeze-branding and radio tagging, which would make it possible for Sea World to avoid harassing orcas and pods that it had previously captured or attempted to capture. Evans also defended the Sea World permit a few days after the MMPA reauthorization hearings, when he appeared before the Senate, which was considering his nomination as the new MMC chairman. Introduced by Bill Lowery, the representative for his district in San Diego, as "one of the world's foremost marine mammalogists and bioacousticians," who had "pioneered the development of radiotelemetry, monitoring the movement of cetaceans," Evans was careful to point out to the Senate committee that he was aware of the potential conflict of interest between the MMC chairmanship and his role as director of HSWRI. To avoid the appearance of a conflict of interest, he had formally recused himself from "acting, advising, recommending or participating with respect to any particular matter where the outcome may have a direct and predictable effect on Hubbs–Sea World Research Institute or Sea World, Inc.," as well as a long list of areas of marine mammal science and policy in which HSWRI conducted research.[52]

Despite Evans's recusal, the first question posed to him by the chairman of the Senate committee, Bob Packwood of Oregon, regarded HSWRI's role in Sea World's controversial killer whale permit application. In his reply Evans explained that he had "no formal relationship with Sea World, Inc., whatsoever" and that HSWRI was "totally independent" of the theme park, but he also admitted that HSWRI had provided advice to Sea World at various points in the development of the permit application, helping to strengthen the scientific aspects of the proposal. In general, Evans added, he had supported keeping cetaceans in captivity for the purpose of scientific research since his experience with Gigi. Even though he had initially been "very outspoken" in his opposition to the capture of the young gray whale, he told Packwood, he had changed his mind because of the scientific information it ultimately produced, including the data resulting from his own radio-tracking study. Sea World's proposed killer whale operation would, he expected, provide even greater benefits to science, especially since Sea World had reshaped its plans in response to criticisms and suggestions from HSWRI, the MMC, and the broader scientific community. The result was a "good research program."[53]

Although the opponents of Sea World's proposal ultimately failed in the legislature just as they had failed in the regulatory process, they eventually succeeded in blocking the capture through the judiciary. On May 1, 1984, the Sierra Club Legal Defense Fund sued NMFS in federal court in Alaska on behalf of Greenpeace, the Southeast Alaska Conservation Council, and a group of tour boat operators, arguing that the permit should not have been granted before an environmental impact statement had been prepared as required by the National Environmental Policy Act. Later that month, Alaska governor Bill Sheffield announced his opposition to the capture, citing opposition from the general public of Alaska, as well as the special concerns of "native people in southeastern Alaska who have strong cultural ties to the orca." A few days after Sheffield's announcement, Greenpeace held demonstrations at Sea World's parks in San Diego and Ohio, handing out leaflets and displaying a large inflatable whale it had first used in its antiwhaling campaign. In June, Sea World announced that it had decided to delay attempting to capture killer whales until the following year. Although Sea World told the press that it had no interest in acting without the approval of the Alaskan public, it was also careful to note that Alaskan opposition had no legal significance, since the MMPA had taken jurisdiction over marine mammals away from the states. Early in 1985, however, a U.S. district court judge in Anchorage ruled in favor of Greenpeace and other environmentalist groups, voiding Sea World's permit on the basis of NMFS's failure to prepare an environmental impact statement for an action—the issuance of the permit—with the potential for significant environmental harm. NMFS's appeal of the ruling dragged the case out for another year and a half, but the Ninth U.S. Circuit Court of Appeals ultimately upheld the district court's decision.[54]

Eventually NMFS would succeed in blocking most legal challenges to MMPA permits under the National Environmental Policy Act by adding a statement in its *Federal Register* notices that it had determined that a particular permit did not require an environmental impact statement. In the short term, however, the door to other challenges to permits on environmental grounds had been opened by the successful case against Sea World's killer whale permit. In April 1986, NMFS submitted a notice to the *Federal Register* that A. Rus Hoelzel, a doctoral student at Cambridge University, had applied for a permit to take up to eighty-six killer whales in the Puget Sound area by harassment, including forty-five by skin biopsy using darts fired from a crossbow. The skin samples would be used to determine the extent of inbreeding in the killer whale population and to test for pollutants, both of which were potential constraints on the health of the population. NMFS issued the permit that August with a number of restrictions. Hoelzel was required to consult and coordinate closely with the NMFS office in Seattle, seeking its authorization at the beginning of each field season and providing a detailed report at the end of each year. He was prohibited from attempting to dart any given orca more than once per day or more than twice throughout the duration of the permit, to actually dart any animal more than once, or to dart more than five animals in the first year of the study. All research activities would be suspended if there was any uncertainty about whether the darting had injured a whale or altered the social relations of a pod, if a dart accidentally struck a nontargeted whale or struck a targeted whale anywhere besides its dorsal midbody region, or if a designated NMFS observer decided that there was a possibility that the research was causing "substantial disruption" to a pod.[55]

In contrast to the Sea World permit, there had been virtually no public discussion of the merits of Hoelzel's proposal. Even though Hoelzel had been conducting observational studies of the killer whales for several years before submitting the darting proposal, Greenpeace and several prominent killer whale researchers seem only to have become aware of the proposal after the permit was issued. In late September 1986, Greenpeace asked NMFS to revoke the permit so that it could be reconsidered with input from its opponents, who included both Balcomb and Spong. While NMFS's Seattle office argued that information about inbreeding and pollutant levels was essential to protecting the whales, Balcomb argued that enough was already known and that the highly likely public backlash against Hoelzel's study might constrain all scientific research on the whales. The small amount of information to be gained from Hoelzel's study, he argued, "doesn't justify the adverse publicity that research in general would get from this invasive technology." Hoelzel told the press that he was surprised by the opposition from Greenpeace, an organization he respected, and noted that researchers had used far more invasive techniques in the past, including capturing orcas to attach tracking devices. He would immediately halt his research, he added, if the whales reacted to darting any differently than they did to being photographed. Comparing his work to the kinds of research that Sea World had proposed in 1983, he said, was "a bit like comparing a ladyfinger with an atom bomb." If anything, the ability to acquire tissue samples by darting "takes away from the argument that you have to restrain an animal to do good science."[56]

When NMFS refused to revoke or reconsider Hoelzel's permit, the Sierra Club Legal Defense Fund filed a suit against NMFS on behalf of Greenpeace in the U.S. district court in Seattle alleging that, as in the case of the Sea World permit that had been definitively invalidated that summer, NMFS had violated the National Environmental Policy Act by failing to prepare an environmental

impact statement before issuing the permit or to provide a reasonable explanation of why a statement was unnecessary. The suit also alleged that NMFS had
violated the MMPA by failing to collect sufficient information to determine
whether the permitted actions were for "bona fide" scientific purposes and consistent with the aims of the MMPA. Despite Hoelzel's claims to the contrary,
Greenpeace representatives told the press, his techniques were far more invasive
than photo-identification. By this time Hoelzel had already returned to Cambridge for the season without having darted any whales. The following spring,
at the beginning of that year's field season for killer whale research, a federal
judge issued a temporary restraining order to prevent Hoelzel from carrying
out his darting study. It was followed by a more formal injunction in early June,
after which Hoelzel gave up on darting wild whales and instead began collecting
samples from dead and captive whales.[57]

A More Streamlined and Efficient Process?

In advance of the hearings on the reauthorization of the MMPA in 1988, Greenpeace and other environmental and animal protection groups mobilized to con-
solidate in legislation the victories they had won in the courts and to contest
what they saw as the progressive weakening of the act through amendments and
regulatory decisions. The cover of the January-February 1988 issue of *Greenpeace* magazine featured a breaching humpback with the tagline "Can We Save
the Marine Mammals?" and the corresponding article claimed that the initially
promising MMPA had since become "riddled with exemptions, loopholes,
amendments and extensions," undermined by lawsuits, and neglected by the
White House. NMFS's permitting process had proved to be "based on politics
and special interest group pressure rather than on the interests of the marine
mammals," even in the case of scientific research. Although NMFS's own regulations required that any proposed research be "bona fide," necessary, and desirable in light of the overarching purpose of the MMPA, the article noted, "the
review process often falls short of these requirements." Dean Wilkinson, Greenpeace's legislative director, was quoted as saying that the process had "become so
routine that only truly outrageous applications are rejected." In subsequent articles Greenpeace urged its members to write to the relevant committees in the
House and Senate in support of various amendments to strengthen the MMPA,
including amendments to tighten federal oversight of scientific research and
public display of marine mammals.[58]

Though the MMC later noted that "considerable attention was given to revising the Act's permit provisions" in the course of the 1988 reauthorization process, the congressional hearings were dominated by the question of whether the MMPA required that the incidental killing of porpoises by the tuna industry be completely eliminated or merely reduced to a level that did not threaten the health of porpoise populations and the ecosystems to which they belonged. In May Evans testified before the House Subcommittee on Fisheries as director of NMFS, a position to which the Reagan administration had appointed him in the fall of 1986. Although the prepared statement he submitted for the record briefly mentioned that "a number of concerns" about the permit process for scientific research and public display had recently been brought to NMFS's attention, Evans's oral statement to the committee did not mention the issue. The closest he came to discussing the MMPA's effect on scientists' choice of research techniques was when Anderson, the original sponsor of the MMPA, asked him to explain why the concept of "optimum sustainable population" built into the law at the recommendation of marine mammalogists had been so difficult to define and implement. Evans replied that it was extremely difficult to obtain even the most rudimentary estimates of population size for most species of cetaceans, with the possible exception of gray whales, pilot whales, and killer whales, which followed predictable migration pathways close to shore—at which point Anderson interrupted: "We take pictures of them as they go by, so there's no problem there."[59]

The scientific research permit process was, however, a significant concern for some of the representatives of animal protection and environmental organizations who testified at the reauthorization hearings. Friends of the Sea Otter was particularly concerned about a proposed amendment authorizing NMFS to issue "emergency permits" to allow scientists to take advantage of "unique research opportunities" without the delay imposed by the standard thirty-day public review period. Such a change would, the organization's representative argued, "create a loophole that is far too large for the limited benefits it could produce." Bo Bricklemeyer, a Greenpeace representative testifying on behalf of a coalition of more than twenty animal protection and environmental groups, focused most of his oral testimony on the tuna-porpoise issue, but he also mentioned that the coalition supported tightening the requirements for scientific research and public display. The basis of that position was a statement prepared by the Animal Protection Institute, which, echoing Wilkinson's statement in *Greenpeace*, claimed that the "permit review process has become so automatic and routine that only when there is a truly outrageous permit application or when there is a threat of litigation will an application be rejected." Out of nearly three hundred research permit applications submitted to the MMC since the MMPA had gone into effect, the statement asserted, only two had been rejected, and though most of the permits involved nonlethal research, NMFS had authorized the killing of more than 27,000 animals. The review process rarely took into account whether the proposed research was "necessary," even in the case of lethal takes, where NMFS's own guidelines suggested that reviewers take into consideration whether the permit furthered "a bona fide and necessary or desirable scientific purpose."[60]

The animal protection and environmental groups were largely successful in their effort to strengthen the MMPA's permitting process. The 1988 reauthorization of the MMPA added new requirements that scientific research on marine mammals be "bona fide" and not unnecessarily duplicative, and it stipulated that permits for lethal research could only be issued if the applicant had demonstrated that there was no reasonable alternative. Permits to take animals from "depleted" populations could now only be issued when the take would benefit the species or the population, and the proposal to allow NMFS to issue "emergency permits" without public review was rejected. The reauthorization also strengthened the requirement that theme parks and aquariums have an educational or research component in order to receive a permit for public display, though it also allowed for the taking of depleted species for captive breeding and other methods of "enhancement" of the stock or the species, a provision the animal protection groups had viewed with skepticism. Greenpeace told its members that it had won much of what it hoped for but that continued vigilance was necessary to ensure that NMFS implemented the amendments, almost all of which it had opposed.[61]

In response to the extensive litigation of the mid-1980s and to the 1988 amendments to the MMPA, NMFS began a comprehensive review of its permit program, including a series of working sessions in which scientists were invited to participate. In the June 1989 issue of the journal *BioScience*, marine mammalogists Katherine Ralls and Robert L. Brownell Jr. urged their colleagues to pay careful attention to these discussions. Echoing Norris, Schevill, and Ray's testimony at the ocean mammal hearings in the fall of 1971, Ralls and Brownell argued that the active management of nature became increasingly necessary as the influence of human activities on the planet grew. Management in turn depended on science, so any restrictions on research, even those intended to protect marine mammals, were ultimately counterproductive. Moreover, they argued, "the potential benefits of basic research are commonly underestimated" in the regulatory process because, as Thomas Kuhn had shown in his *Structure of Scientific Revolutions*, science was "a uniquely powerful technique for producing surprises," inherently unpredictable and therefore inherently impossible to regulate on the basis of its possible future benefits. Even the most apparently trivial or theoretical research might turn out to have profound implications for conservation. Despite the good intentions of the people who had proposed the 1988 amendments to the permit process, Ralls and Brownell warned, the new requirements "could be implemented in such a way as to impair the freedom of scientists and retard the acquisition of knowledge needed for sound conservation strategies." Restrictions on marine mammal research were "already stringent," they concluded, and "adding even more layers of bureaucracy may easily do more to inhibit research than to protect species."[62]

While some biologists fought to limit the government regulation of research, others focused their energy on the development of new techniques that would provide detailed information about the lives of marine mammals without raising significant concerns among regulators or animal protectionists. Since Hoelzel's failed attempt to collect biopsies from killer whales in 1986 and 1987, no invasive techniques had been used to study the orcas of the Pacific Northwest. In the spring of 1990, the participants in a Workshop on Comparative Behavioral Ecology of Killer Whales held at Friday Harbor, Washington, issued a report suggesting that radio tags and biopsy darts could provide useful information about the population, but as the *Marine Mammal News* reported, the authors of the report "took note of the sensitive nature of electronically marking individual whales." The following year Baird, then a graduate student at Simon Fraser University in British Columbia, developed a method of radio-tagging orcas that eliminated two of the most sensitive aspects of previous tagging attempts: the need to temporarily capture the whale and the attachment of the tags with bolts that penetrated the dorsal fin. With the help of whale researcher Jeffrey Goodyear, who had first tracked a humpback whale using what he called a "remora" tag a decade earlier, Baird developed a tag that could be attached to free-swimming whales by suction cup. Baird attached his first tag in 1991 using a long pole; in 1993 he had more success with a crossbow, attaching seven tags in sixty attempts. The tags remained attached for hours or days at most, but they still provided valuable data about movement and diving behavior. In his doctoral thesis, Baird left out the controversial history of tagging, instead explaining the almost complete absence of previous radio-tracking studies of killer whales in terms of the research priorities of the previous generation of researchers and the geographical peculiarities of the Puget Sound area, which made it easy to observe the whales visually. The development of a "non-intrusive method of tag attachment" was important, he wrote, because it reduced the possibility of influencing the whale's behavior and thereby biasing the data collected.[63]

By the end of 1992, a year before the MMPA's authorization was due to expire, NMFS had not yet managed to implement all of the regulatory changes required by the amendments of 1988. It was still trying to determine how to evaluate the research and educational programs of public display institutions, how to decide whether proposed research was "bona fide" and not unnecessarily duplicative of previous studies, and under what conditions it was appropriate to issue permits for the "enhancement" of depleted populations. It was also struggling to manage the increasing demand for permits to study certain species, particularly humpback whales and killer whales off the coasts of Hawaii and Alaska, for which there were sixteen permits active in 1992 authorizing the take of up to 6,670 humpbacks and 1,030 killer whales for "photo-identification, radio-tagging, acoustic, and genetic studies and aerial and vessel surveys." To coordinate these increasingly overlapping and occasionally conflicting permits and to determine whether the permitted activities in the aggregate might be having damaging effects that none of them would have had alone, NMFS convened a series of meetings in 1992 among its staff, permit holders, and scientific advisors in Washington, DC, and Seattle. When the scientific advisory committee held its final meeting on the issue that fall, it concluded that many of the permit applications being received by NMFS were incomplete and should be returned for more information that would allow NMFS "to determine whether the research is *bona fide*, not unnecessarily duplicative, and humane."[64]

In late July 1993, as part of a series of hearings on the reauthorization of the MMPA, the Senate Committee on Science, Commerce, and Transportation met to hear testimony on proposed changes to the MMPA's permitting process for public display and scientific research, with Senator John Kerry of Massachusetts presiding. In his opening statement, Kerry noted that many Americans had been touched and entertained that summer by *Free Willy*, a movie about "a troubled young boy who befriends a whale that is taken from its ocean habitat and put into an aquarium." Much as the institutions that put marine mammals on display claimed to do, Kerry noted, the film had heightened the public's

Robin Baird attached a radio tag and time-depth recorder to this 13-year-old male killer whale, known as L58, on June 24, 1993, near Race Rocks off the southeastern tip of Vancouver Island. The tag's suction cup held for 8 hours and 24 minutes, longer than any of the other six tags Baird succeeding in attaching that season. (Courtesy of Robin W. Baird/www.cascadiaresearch.org)

appreciation of the animals and of the ecosystems of which they were a part, while also raising questions "about the habitat in which these mammals are kept, about the permitting process, about the conditions in which they live, and about the treatment which they are afforded."[65]

While representatives of the public display industry at the hearings sought to portray their institutions as primarily sites of science and education, representatives of animal protection groups emphasized their profit-making motives and questioned the necessity of research that involved harming marine mammals. John Grandy, the vice president for wildlife and habitat protection of the Humane Society of the United States, and Naomi Rose, the organization's marine mammal scientist, who had been studying killer whales off the coast of British Columbia for the previous seven years, both testified that they believed the value of research on captive cetaceans had been exaggerated and that lethal research on marine mammals in the wild had been rendered largely unnecessary by the development of noninvasive techniques. Invasive and lethal research should only be permitted, they argued, when it would directly benefit the species in the wild. Grandy told the committee that "given more and more of what we know today and what is available to us through video and other things shows less and less value, candidly, to doing studies in animals that are isolated in these small, incredibly tiny aquariums." When Kerry asked Grady whether he was "suggesting that the commercial enterprise is at the foundation of the effort, and that the research has become the veneer to justify it," Grady replied that he would not have put it in exactly those words but that he "could not have said it better" himself.[66]

Kerry also challenged representatives of the public display industry to defend the claim that research on captive animals ultimately benefited marine mammals in the wild. It seemed likely, Kerry told Jim McBain, the corporate director of veterinary medicine for Sea World, that captive animals adapted to their artificial habitat and became "different from the animals that came out of the ocean"; scientists who studied them were "not really studying the original animal any more." McBain replied that he believed captive animals changed little in captivity, except in the unfortunate case in which they were held in isolation, as the orca featured in *Free Willy* had been. Representatives of the public display industry also argued that field research was not sufficient by itself, despite the advances in methodology that Grandy had described. Michael Hutchins, director of conservation and science for the American Association of Zoological Parks and Aquariums, told Kerry that information about the reproductive biology of marine mammals, for example, would be both more difficult and more inhumane to collect in the field than in captivity. Moreover, Hutchins added, research on captive animals was essential for developing new techniques of field research: for example, "methods used to identify animals or to track and monitor them using biotelemetry or radiotelemetry are first tested on animals in captivity. And then those technologies, once they have been developed, can then be transferred to the field."[67]

Several witnesses at the hearing argued that, regardless of the validity of critiques of certain research methods, the progress of marine mammalogy had been hindered by the MMPA's increasingly burdensome requirements. Robert Hofman, the director of the MMC's scientific program, told the committee that in the past few years, "scientists have complained that it takes longer and more information is required than should be necessary to obtain permits for research" and "that permits often contain unnecessary and burdensome reporting requirements." When Kerry asked Hofman about the proposal of animal protection groups to prohibit lethal and invasive research that did not directly benefit the species concerned, Hofman referred to Ralls and Brownell's *BioScience* article of 1989, which had argued, he told Kerry, that "it is very, very, very difficult—it is impossible—to predict beforehand how the results of certain kinds of research may contribute to or benefit a species." Hofman's main example of the unpredictability of research was photo-identification of killer whales. In 1976, he told Kerry, its value had been "controversial" in the scientific community; many had "felt that it was a bogus scientific practice, and would never result or provide anything that was useful. . . . In fact, many thought it was plain harassment, and it was harmful." Now, twenty years later, it had become "one of the most important techniques for studying many species of animals." When Kerry asked Hofman to explain for the record the benefits "of the intrusiveness that we do permit into the habitat and the life of marine mammals," Hoffman replied that scientific research had been essential to solving many of the issues that had inspired the original passage of the MMPA in 1972, such as the "ineffective regulation of commercial whaling, the incidental take of porpoise in the yellowfin tuna purse sein[e] fishery, and the clubbing of baby harp seals in the North Atlantic." Now new forms of invasive research, such as satellite tracking, were helping scientists and wildlife managers make "sound conservation decisions." In fact, Hofman said, it was his opinion "that the problem that we are having right now is overregulation rather than not enough regulation."[68]

Despite the efforts of animal protectionists to build on the achievements of 1988, the 1994 amendments to the MMPA significantly loosened restrictions on permits for scientific research and public display. For research permits, it eliminated the "nonduplicative" stipulation that had been imposed in 1988, allowed the thirty-day public review period to be waived for "emergency" situations, and established a new category of "minor" amendments to permits that would not be subject to public review. Rather than being left to NMFS or the Fish and Wildlife Service to determine, the meaning of "bona fide" research was now defined statutorily as research that was likely to be published in a refereed scientific journal, to contribute to basic knowledge of biology or ecology, or to contribute to conservation. Most important, for scientific research, the 1994 amendments established two levels of "take." Level A was defined as "any act of pursuit, torment, or annoyance which has the potential to injure a marine mammal or marine mammal stock in the wild," while Level B was defined as "any act of pursuit, torment, or annoyance which has the potential to disturb a marine mammal or marine mammal stock in the wild by causing disruption of behavior patterns including, but not limited to, migration, breathing, nursing, breeding, feeding, or sheltering." Although researchers would still have to notify NMFS or the Fish and Wildlife Service if they intended to conduct research involving only Level B takes, such research would no longer have to go through the full review process, instead being covered under a "general authorization" for activities such as "photo-identification studies, behavioral observations, and vessel and aerial population studies."[69]

Instead of inspiring heightened protection for marine mammals by the federal government, the public concern generated by *Free Willy* was channeled into private efforts to return the movie's orca star, Keiko, to the wild. Efforts to rehabilitate and release Keiko, a whale originally captured in Icelandic waters who was being kept in poor conditions in a Mexican theme park at the time of the filming, began almost as soon as the movie was released. The producers of the film had arranged for a toll-free telephone line that could be called by viewers interested in protecting killer whales, which was to be managed by the Earth Island Institute. David Phillips, the director of the Earth Island Institute, later recalled that he had prepared for as many as 25,000 calls to be generated over the course of the film's theatrical run, but the eventual number was closer to 300,000. Many callers were more interested in saving Keiko than they were in saving orcas or whales in general. According to nature writer Kenneth Brower, Phillips was initially ambivalent about getting involved in the effort to free Keiko, since the Earth Island Institute's focus had always been on saving species and ecosystems, not on saving individual animals, but soon changed his mind when he saw the enormous interest in killer whales that *Free Willy* and Keiko had generated.[70]

By the 1990s animal protectionists and killer whale activists had won their campaign to end the capture of killer whales in American and Canadian waters and had begun to focus their attention on the release of captive whales, particularly those captured in the Pacific Northwest, such as Corky at Sea World's San Diego park and Lolita at the Miami Seaquarium, whose original pods were well known after several decades of close study. Unlike these whales, Keiko came from an unknown pod and had been captured when very young, making him a more challenging candidate for reintroduction to the wild. No other whale, however, had attracted the attention and financial contributions that Keiko had. By 1995, the newly established Free Willy/Keiko Foundation had reached agreements with the Reino Aventura theme park to sell Keiko and with the Oregon Coast Aquarium to build a new holding facility for him. The aquarium's application for a permit to import Keiko for public display was published in the *Federal Register* on August 14, 1995, NMFS issued the permit in late September, and Keiko was transferred to his new tank at the Oregon Coast Aquarium early the following year.[71]

The Free Willy/Keiko Foundation intended Keiko's move to the Oregon Coast Aquarium to be a temporary stop on his way to freedom, but the possibility and advisability of releasing him was widely questioned. Although the 1994 amendments to the MMPA had stripped NMFS of its authority over captive marine mammals once they were at a permitted facility, NMFS told the Free Willy/Keiko Foundation and the Oregon Coast Aquarium that they would need an additional scientific research permit before they could legally release Keiko. Such a release could result in a "take," NMFS argued, either by injuring the released animal or by allowing the released animal to pass diseases to or otherwise harm the wild marine mammals with whom it came into contact. No release permit, NMFS assured the press, would be issued without the approval of both NMFS and the MMC, which would be contingent on a plan for monitoring the released orca for at least a year after release. The MMC further specified that its approval for the release of Keiko or any other captive cetacean would not be forthcoming until "objective, generally accepted criteria had been developed for judging when release is appropriate."[72]

That NMFS would use its authority to discipline those who released captive marine mammals without its approval was demonstrated following the release of two dolphins without release permits in 1996. The dolphins had been acquired from the U.S. Navy Marine Mammal Program by the Sugarloaf Dolphin Sanctuary in Florida with the assistance of Ric O'Barry, the former trainer of television's "Flipper" and a prominent advocate for the release of captive cetaceans. After O'Barry and a colleague released the dolphins in May 1996, they were quickly recaptured by the navy and NMFS, which claimed that the dolphins had been released before they were able to feed themselves. NMFS filed charges early in 1998, and a year and a half later O'Barry and his colleague were found guilty of violating the MMPA. They were fined $40,000 for illegally "taking" marine mammals, while the Sugarloaf Dolphin Sanctuary was fined $19,500 for failing to notify NMFS of the proposed release.[73]

The MMC and NMFS consistently argued that no responsible release could be conducted without a means of long-term monitoring of the released animal, which would probably require some kind of durable radio tag. As if to prepare Keiko's fans for the need for such a tag, radio tracking was given a central role in *Free Willy 3: The Rescue*, which appeared in theaters in 1997. Like *Free Willy 2: The Adventure Home*, which had been released two years earlier, the movie used animatronics, computer graphics, and footage of wild whales to simulate Keiko, who by this time was being rehabilitated and displayed at the Oregon Coast Aquarium. In *Free Willy 3*, Willy is living somewhere in the Pacific Northwest, and Jesse, the boy who had befriended him as a child, is now a teenager working as a researcher for a study of the causes of a mysterious decline in the population of killer whales, which is eventually revealed to be the result of illegal whaling. The study relies on satellite tags attached to Willy and his pod with suction cups, which, Jesse is told, will make it possible "to track our whale anywhere on the planet, twenty-four hours a day." Throughout *Free Willy 3*, satellite tags serve as the mechanism by which killer whales, whalers, and antiwhalers are brought together in the narrative, ultimately to the benefit of the whales. The tags that Jesse has placed on Willy and his pregnant mate and on the whalers' boat eventually make it possible for him to prevent the whalers from killing Willy's pod.[74]

In the fall of 1997, the Free Willy/Keiko Foundation began discussions with NMFS about relocating Keiko to a sea pen in the North Atlantic, where he would continue to be rehabilitated and eventually released, perhaps to rejoin his original pod. The Oregon Coast Aquarium, however, which had experienced a dramatic boost in visitors since acquiring the celebrity cetacean, argued that

Keiko, the killer whale star of the movie *Free Willy*, was kept in a specially built tank at the Oregon Coast Aquarium before being transferred to a sea pen in Iceland in September 1998. (Public Domain, U.S. Department of Defense)

Keiko was not ready for release and perhaps never would be. Despite the aquarium's efforts, NMFS transferred the public display permit for Keiko to the Free Willy/Keiko Foundation in the spring of 1998, opening the door for Keiko's transportation to Iceland, where a suitable site for a sea pen had been found. On August 17, 1998, after an expert committee had reviewed Keiko's health and the plans for his rehabilitation in Iceland, NMFS told the Free Willy/Keiko Foundation that it had satisfied its requirements for exporting Keiko, although it also continued to stress that an additional scientific research permit would be required before Keiko could be legally released. A few weeks later, Keiko was flown in a U.S. Air Force cargo plane from Oregon to Heimaey, the largest of Iceland's Westmann Islands, where he was trucked to his sea pen in Klettsvik Cove.[75]

While some of the funding for Keiko's rehabilitation came from small donors, the bulk of the multimillion-dollar effort was funded by telecommunications billionaire Craig McCaw. In 1999, his ex-wife, Wendy McCaw, donated more than $150,000 to the National Marine Mammal Laboratory to develop a satellite tag for Keiko. Doug DeMaster, the director of NMML, told journalist

Katy Muldoon of the *Oregonian* newspaper that the grant was "a tremendous opportunity for us to develop a satellite tag suitable for small cetaceans all over the waters of the United States." For Keiko the tag would, in Muldoon's words, provide "something of a security blanket," making it possible for the rehabilitation team to take him on guided "walks" outside of his pen without worrying that he might be lost, much as the Deep Ops team had used radio tags to keep track of Ahab and Ishmael thirty years earlier. Jeff Foster, who had previously participated in the capture of killer whales in Icelandic waters and was now leading the rehabilitation effort, told Muldoon that electronic tagging of cetaceans was still far from routine, however, as had been demonstrated by Sea World's recent attempt to track a young gray whale named J.J. The whale had been rehabilitated at the marine park and then released with a radio tag, but the tag had detached within days, most likely because J.J. had rubbed it off against the sea floor, just as Gigi had probably done with Evans's radio tag in 1972.[76]

The small-cetacean radio-tagging team at NMML was led by Brad Hanson, who had been conducting a detailed bioengineering study of methods of attaching tags to the dorsal fins of small cetaceans for his doctorate at the University of Washington. As undergraduate research assistants, Hanson and Foster had helped Erickson radio-tag the killer whales captured in the Budd Inlet in 1976. Wendy McCaw's grant helped Hanson and his colleagues at NMML expand an existing project focused on Dall's porpoises and belugas. These species were both easier and less controversial to handle than orcas, although permits from NMFS and extensive discussions with local environmental groups and, in the case of belugas, native Alaskan groups were still necessary. Their first real successes came in 1999, when they satellite-tracked a beluga tagged in the Cook Inlet near Anchorage, Alaska, for nearly four months. By the spring of 2000, Hanson's team had succeeded in tracking one Dall's porpoise for nearly a year. That May, Hanson traveled to Iceland to fit two tags to Keiko's dorsal fin. One was a conventional radio tag that would allow Keiko to be located continuously in the field; the other was a satellite tag that would transmit his location and diving behavior several times a day. Keiko's handlers had recently intensified his training in expectation of gradually introducing him to the open ocean and to the pods of wild killer whales feeding on herring not far from his pen over the course of the summer, and Ocean Futures, the environmental organization that had taken responsibility for Keiko after he was moved to Iceland, had submitted a permit application for Keiko's release to Iceland's Animal Welfare Board, copies of which were sent to NMFS and the MMC.[77]

With Hanson's tags attached to his dorsal fin, Keiko was taken out for his first "walk" outside Klettsvik Cove on May 25, 2000. A few weeks later, the MMC sent its comments on the release protocol to Ocean Futures, noting that it still had questions about how Keiko would be monitored after his release, including "what has been or will be done to ensure that the tag cannot be rubbed off" and "how Keiko will be monitored after release in light of potentially severe weather and sea state conditions around Iceland." Around the time the MMC sent its letter, Foster's team decided to remove Keiko's tag after noticing the appearance of a gap between the tag and his skin. The tag was returned to Hanson to be repaired and refitted using a model of Keiko's dorsal fin, a two-week-long process during which period Keiko's "walks" had to be temporarily suspended. The adjusted tag was reattached to Keiko in early July, and a few weeks later Ocean Futures informed the MMC that it expected that the improved tag would allow Keiko to be tracked for as long as a year.[78]

Though Keiko was taken on a number of "walks" in the summer and fall of 2000, he barely interacted with the wild whales he encountered. At the end of the summer, with Keiko's window of opportunity to join a wild killer whale pod closing for the season, the rehabilitation team led Keiko out to the few remaining wild killer whales and then slipped away, leaving him to fend for himself for a week. Although Keiko's radio tag made it possible for the team to monitor him closely by helicopter, there was no evidence that Keiko was feeding himself during this period, and he continued to interact little with the wild whales. The results of similar efforts during the following summer were equally disappointing to those who had hoped that Keiko would leap at his first chance at freedom. By the end of 2001, after Craig McCaw decided to significantly reduce his support for the project, Ocean Futures concluded that Keiko's release might take longer than envisioned and began to consider moving him to a less expensive, more accessible, and possibly permanent sea pen elsewhere in Iceland.[79]

In the summer of 2002, however, the Humane Society of the United States took control of Keiko's rehabilitation in partnership with a revitalized Free Willy/Keiko Foundation, dismissing Foster and his team and replacing them with a new set of trainers led by Charles Vinick and Colin Baird, who HSUS expected to pursue a more aggressive release strategy. That spring Hoelzel, who had eventually succeeded in obtaining biopsy samples from live killer whales in the Pacific Northwest and elsewhere, reported that genetic analysis of samples that he had acquired from twenty-nine wild Icelandic killer whales revealed that Keiko had been swimming among cousins the previous two summers, though none of the whales were likely to have been members of his original pod. A new satellite tag arrived in Iceland and was fitted to Keiko under the direction of former Sea World veterinarian Cornell in early July. On July 7, a day after the attachment of the new tag, Keiko was taken out on his first "walk" of the season to encounter wild killer whales. On July 15, 2002, after spending five days with the wild killer whales, Keiko swam back to the sea pen of his own accord, where he was fed and the batteries for his radio and satellite tags were replaced. After he was led back out, Keiko immediately separated from the boat to rejoin the wild whales, eventually following them as they swam eastward toward the Faroe Islands. The rehabilitation team followed him closely with the help of the radio tag until the end of the month, when a storm forced the tracking ship back to port.[80]

Although Keiko had not been formally "released," he seemed to have finally seized the opportunity to live independently of his caretakers. On August 2, 2002, HSUS issued a press release that asked, "So is Keiko free?" Its answer: "Almost." A week later data from Keiko's satellite tag placed him near the Faroe Islands, about 165 miles away from his pen in Klettsvik Cove. Vinick and Baird flew with radio-tracking gear to the Faroe Islands, where they were met a few days later by the tracking boat. Their goal was to observe Keiko in the flesh to determine whether he was feeding, in good health, and in the company of other whales, but even with the help of the satellite locations and the conventional radio tag, they were unable to locate him. Nonetheless, on August 16, after Keiko had spent forty-one days outside of his sea pen—not counting his voluntary return in mid-July—HSUS issued another a press release suggesting that his "quest for freedom may be fulfilled" and quoting HSUS marine mammal expert Naomi Rose as saying that Keiko's behavior gave "good reason to hope that he is on the verge of true independence."[81]

For the next two weeks the satellite tag indicated that Keiko was swimming long distances eastward each day, which HSUS took as a sign that he remained in good health. Foster and seven of Keiko's former handlers from the period when Ocean Futures had been in charge of the project, however, believed that HSUS was jumping to conclusions in its eagerness to claim that Keiko was "free." In late August 2002, the group wrote to NMFS to express their concern that Keiko was not ready for release and that he was spending time in an area not known to have other killer whales. Satellite monitoring, they insisted, was incapable of proving that Keiko was healthy; they recommended that the team establish visual contact and lead Keiko back into captivity if necessary. Some

Colin Baird and the Keiko team attached new tags to the killer whale in early July 2002. The satellite tag, designed by Brad Hanson at the National Marine Mammal Laboratory in Seattle, was larger than the conventional radio tag and was placed lower on the dorsal fin. In August 2002 the satellite tag allowed the team to track Keiko's voyage from Iceland to Norway and to monitor his diving behavior. (Courtesy of Fernando Ugarte)

members of the HSUS team, in contrast, believed that Keiko's hesitancy to engage with wild whales during his first two years in Iceland had been the result of coddling by Foster's team. On August 28, 2002, in a teleconference with the MMC and NMFS, the HSUS project leaders expressed optimism about Keiko's health based on the long distances he was swimming each day and the pattern of diving behavior reported by the satellite tag, which suggested that he was foraging for fish. According to a later report by the MMC, "the project leaders concluded that Keiko was successfully adjusting to life in the wild and that continued satellite tracking was not necessary."[82]

The day after the HSUS team had come to this hopeful conclusion, Keiko was sighted just off the coast of Norway, where he began following fishing boats and interacting with local people and, soon, tourists who had come to see the celebrity whale in the flesh. The HSUS team flew to Norway and led Keiko by boat northward to an isolated fjord, where he was confined within a quickly constructed sea pen to protect him from dangerous interactions with boats. Keiko's return to humanity after his brief and contested period of independence effectively marked the end of efforts to return him to the wild. In September 2002, the Miami Seaquarium, the home of Lolita, submitted an application to NMFS to import Keiko from Norway and told the press that it had "offered to rescue Keiko" from his current situation. NMFS rejected the application on the basis that the Norwegian government had not yet granted permission for Keiko's capture or export. Keiko spent the next year in his sea pen in Taknes Fjord under the close observation of a small team of caretakers, eventually dying on December 12, 2003, from acute pneumonia. After Keiko's death, Vinick told the press that he "was not truly a wild whale. He was not truly a captive whale. He was somewhere in between."[83]

Keiko's position between captive and wild, however atypical or even unique it may have seemed, was an increasingly common one for killer whales and other wildlife at the end of the twentieth century. Most, of course, were not captives in theme parks or candidates for release from sea pens, but virtually all lived in environments profoundly transformed by human activity. In 2005, the population of southern resident killer whales in Washington State and British Columbia was listed as endangered under the Endangered Species Act because of population declines widely attributed to the collapse of the salmon runs in the Pacific Northwest. The recovery plan finalized by NMFS several years later noted that the development of "noninvasive techniques" for studying the whales was desirable but that telemetry and other tagging studies were essential to answering long-standing questions about the population. In a world where whales and other wildlife were always already influenced by humans, the caretaking or stewardship vision that Norris, Schevill, and Ray had presented at the 1971 ocean mammal hearings seemed increasingly compelling even to many animal protectionists and wilderness advocates. In such a world, the argument that Scheffer had presented against Sea World's proposed capture operation in the early 1980s—that a killer whale in a pool was not a killer whale—no longer made sense. If "pool" is taken as a metaphor for a human-constructed environment, all killer whales were in pools, and they all could potentially benefit from the kind of continuous surveillance and intervention that radio tags made possible.[84]

New Connections

From the 1960s to the 1990s, disputes over wildlife radiotelemetry often pitted scientists against wilderness activists, animal protectionists, and others claiming to speak for animals and for the public. At the end of the twentieth century, radio tracking also became a means for wildlife biologists to establish connections between the animals they studied and the mass audiences whose support was necessary for effective conservation. In 1996, for example, ornithologist Dave Anderson of Wake Forest University received funding from the National Science Foundation for what he eventually called the Albatross Project. Although the scientific goal of the project was to study the foraging behavior of albatrosses using tags that transmitted location data via the Argos satellite system, a technique pioneered by French biologists Henri Weimerskirch and Pierre Jouventin in the late 1980s, the project also included a significant public outreach and education component. Using e-mail and the World Wide Web, the locations of the satellite-tagged birds would be transmitted to elementary-school classrooms and to the public almost as soon as they were received by the researchers. By June 1998, six months after Anderson's team attached its first tag to an albatross, Anderson estimated that 10,000 fifth- and sixth-grade students had received regular updates on the movements of individual albatrosses. Though targeted primarily toward schoolchildren, the Albatross Project was open to the public; one newspaper article published early in 1998 invited "readers and nonscientists" to "join the action" by visiting the Web site or subscribing to e-mail updates.[1]

Anderson's Albatross Project was only one of a number of education and outreach efforts that took advantage of technical advances in satellite wildlife telemetry and the growth of the Internet in the 1990s. In the summer of 1998, Collecte Localisation Satellites, the French company responsible for the Argos system, dedicated an issue of its newsletter to the rise of the World Wide Web as a means of distributing tracking data to scientists and sharing results with the public. Although most of the tags tracked by the Argos system were used by meteorologists and oceanographers to collect environmental data, animal tracking received the bulk of the attention. That fall, when *Audubon* magazine published an article on how its readers could "get involved in tracking animal movements," it listed albatrosses, cranes, eagles, elephants, geese, mallard ducks, marine mammals, manatees, and sea turtles as types of animals whose movements were already being broadcast over the Internet. These outreach efforts turned the logic of radio tracking as a pathway to professional authority for wildlife biologists on its head. Instead of being seen as experts whose technologically mediated intimacy with wild animals gave them authority to speak on their behalf, scientists could now be seen as mediators of a kind of virtual intimacy between individual animals and mass audiences, or even as audiences themselves. As Anderson explained to one reporter, what made the Albatross Project different from the usual methods of science education or communication was that "there's no filter." Scientists and schoolchildren were connected to radio-tagged animals in exactly the same way.[2]

In 2002, the marine biologist and conservationist Carl Safina dramatized the findings of Anderson and other albatross researchers in his book *Eye of the Albatross: Visions of Hope and Survival*, which won the John Burroughs Award for natural history writing the following year. Safina centered his narrative on an albatross he called Amelia, a member of the Laysan albatross species that had been satellite-tracked by Anderson's group and by a second research group based at the University of California, Santa Cruz. Once Amelia's satellite tag had been attached, the human observer became essentially passive; it was Amelia herself, Safina writes, who would "draw me a map of her world." As the title of the book suggests, Safina's story is told largely from Amelia's perspective, from which humans appear as the source of a variety of mysterious threats to her survival and that of her offspring. Seeing the sea through Amelia's eyes, readers of *Eye of the Albatross* are meant to sympathize with her plight, to recognize their responsibility, and ultimately to take action to save not just albatrosses but also the ocean as a whole. Safina is optimistic about the possibility of such cross-species sympathy. As a result of his research on albatrosses, he writes, he "began not just seeing the world albatrosses encounter, but—in a subtle shift of perception—seeing the world as an albatross encounters it." This transformation of perception was made possible, he explains, by scientists who were aware of and constantly trying to reduce the risks posed by their research procedures, including tagging, to the birds they studied. Those risks were justified, Safina suggests, because only science has the capacity "to draw out what the animals cannot tell us. To give words to the wordless, and voice to the voiceless, so that we can try to reach the ones among *us* who have so far been beyond words."[3]

Albatross researchers and conservationists had realized in the 1990s that the dramatic growth of longline fisheries in the preceding decades threatened the survival of albatrosses, which often became caught on the baited hooks of the lines as they were unspooled. Satellite telemetry of albatrosses promised to identify the areas of the ocean that were most vital to the survival of the far-ranging birds as well as to provide a media-friendly way of visualizing the lives of animals who sometimes spent years at sea without touching land. At a conference on albatrosses held in Tasmania in 1997, British ornithologist John Croxall, who had begun satellite-tracking albatrosses soon after Weimerskirch and Jouventin's initial study, told his colleagues that he believed the species could serve as "the symbol and the spirit of the status of the global marine system." Anderson's Albatross Project, Safina's *Eye of the Albatross*, and a flood of newspaper and magazine articles over the next several years attempted to convince audiences of the magnificence and the imperiled status of the various species of albatrosses, often relying heavily on the results of satellite telemetry to do so. Perhaps the most unusual attempt to thrust albatrosses to the forefront of public concern came in the spring of 2004, when the British betting company Ladbrokes partnered with the Tasmanian government to raise money for albatross research and conservation. The partnership was the idea of Tim Nevard, an Australian conservationist, who proposed giving Ladbrokes' millions of online gamblers the opportunity to place bets on which of a number of satellite-tagged shy albatrosses would first complete their migration from breeding colonies in Tasmania to feeding grounds off the coast of South Africa. Each bird was given a name and assigned a celebrity "owner"; in Ladbrokes' promotional materials, the satellite tags were described as "jockeys" and the scientists who had attached them as "trainers." While gamblers could check the regularly updated progress of their birds online, the company also offered a separate, gambling-free Web site where schoolchildren could plot the birds' movements for strictly educational purposes. All of Ladbrokes' proceeds from the "Big Bird Race" were to be donated to albatross conservation efforts.[4]

The idea that new media technologies had made it possible for wild animals to tell their own stories directly to mass audiences and that such connections might inspire a popular movement to protect wildlife was not a new one. Over the course of the twentieth century, it had been one of the guiding tropes of wildlife documentary films, in which the role of the human filmmaker in creating, recording, and editing wildlife scenes was often effaced, along with the role of humans in shaping the landscapes in which wild animals lived. That trope in turn drew on the genre of realistic wild animal story pioneered by Canadian nature writer Ernest Thompson Seton and others in the late nineteenth century and continued by books such as *Eye of the Albatross*. In the 1990s, however, this trope was given new life by the development of wildlife radio tags that could record audio and video. One of the most prominent of these new tags was the Crittercam tag, originally developed by marine biologist Greg Marshall in the late 1980s. Over the next several decades, with funding from the National Geographic Society, Marshall refined the design of his tag and used it to record "animal's-eye" footage from a variety of marine and terrestrial species. The work of Marshall's Crittercam team at National Geographic was the subject of a one-season *Crittercam* television series, and the team provided footage for a number of National Geographic documentaries, for the weekly *Wild Chronicles* series, and for a large exhibit at the society's headquarters in Washington, DC. In the fall of 2007, National Geographic hosted a symposium on animal-borne imaging that included events for students, teachers, and the public. Regardless of the context in which it was presented, Crittercam footage was almost always accompanied by an account of the technical work of adapting the tag to each new species, including discussions of the ethical and scientific reasons for developing tags that had a minimal effect on the animals to which they were attached. In contrast, much of National Geographic's promotional material elided the technical work that made Crittercam possible and instead aimed to create the illusion that, with the aid of advanced technology, humans had now been granted the ability to see the world through the eyes of a whale, a shark, or a seal.[5]

When Dwain Warner and William Cochran first proposed radio-tracking birds by satellite to the National Aeronautics and Space Administration in 1962, they identified albatrosses as among the most promising candidates for tracking. The birds were large enough to carry the tags and could easily be accessed from the Midway Island naval base, and despite years of research on their behavior at breeding colonies, almost nothing was known about their long-distance movements. Warner and Cochran had little idea that it would take more than twenty-five years for their vision to become a reality or that, once the technical barrier to the satellite tracking of birds had fallen, it would not merely be a means of collecting otherwise unattainable data on wild animals. It would also be a tool for educating schoolchildren, inspiring conservation efforts, raising funds for scientific research, and even burnishing the environmental reputation of an online gambling company. In addition to producing representations of wild animals, it would be a means of establishing connections to them. Although Warner ceased his work with wildlife telemetry long before his death in 2005, Cochran remained actively engaged in the field into the twenty-first century, collaborating with ornithologist Martin Wikelski and engineer George Swenson, his former boss at the University of Illinois ionospheric research laboratory, to propose a new satellite instrument for tracking the smallest of birds and other animals unable to carry existing Argos tags. In 2007, when Wikelski and a group of colleagues who had united under the banner of International Cooperation for Animal Research Using Space (ICARUS) published a paper outlining the proposal, they noted that "tracking animals over large temporal and spatial scales has revealed spectacular biological information" but that existing satellite tags remained too large for an estimated 81 percent of bird species and nearly 67 percent of mammal species. By making it possible to study these species, a small-animal satellite tracking system would help biologists answer "a number of age-old questions about the natural world." If such a system is ever built, it will almost certainly raise anew all of the complex questions that the radio tracking of wild animals has now been raising for more than half a century.[6]

Abbreviations

Buechner Papers — Helmut K. Buechner Papers, Record Unit 7279, Smithsonian Institution Archives, Washington, DC

Cedar Creek Files — Miscellaneous Files, Cedar Creek Natural History Area, Bethel, MN

Cedar Creek Record — Cedar Creek Field Biology Program Records, Collection #238, University of Minnesota Archives, Elmer L. Andersen Library, Minneapolis, MN

Chapman Papers — Douglas G. Chapman Papers, 1949–1988, University of Washington Libraries, Seattle, WA

Cochran Interview — William W. Cochran, telephone interview by the author, 20 Aug. 2004

Cochran and Swenson Interview — William W. Cochran and George Swenson, interview by the author, Champaign, IL, 20 Aug. 2006

Dean Interview — Frederick C. Dean, telephone interview by the author, 9 Mar. 2009

FWS Wildlife Research Files — Bureau of Sport Fisheries and Wildlife, Division of Wildlife Research, Correspondence Files, 1958–1961, Entry 312, Record Group 22, Records of the U.S. Fish and Wildlife Service, National Archives and Records Administration, College Park, MD

Hubbs Papers — Carl Leavitt Hubbs Papers, MC 5, Scripps Institution of Oceanography, La Jolla, CA

Kenward Interview — Robert E. Kenward, interview by the author, Wareham, Dorset, United Kingdom, 7 June 2005

Kuechle Interview

Valerian B. "Larry" Kuechle, interview by the author, Bethel, MN, 10 Aug. 2004

Leopold Papers

A. Starker Leopold Papers, MSS 81/61c, Bancroft Library, University of California, Berkeley

Marshall Papers

William H. Marshall Papers, 1947–1982, University of Minnesota Archives, Elmer L. Andersen Library, Minneapolis, MN

Mech Interview

L. David Mech, interview by the author, St. Paul, MN, 17 Aug. 2004

Mishra Interview

Hemanta R. Mishra, interview by the author, Vienna, VA, 15 Jan. 2009

MN Game and Fish Research Records

Minnesota Conservation Department Game and Fish Division, Research and Planning Section, Administrative Files, 1957–1966, Minnesota Historical Society, St. Paul, MN

Murie (Adolph) Files

Murie Family Papers, Series I: Adolph Murie Files, 1834–1982, American Heritage Center, University of Wyoming, Laramie

Murie (Olaus) Papers

Olaus J. Murie Papers, CONS 90, Denver Public Library, Denver, CO

NPS AK Task Force Files

Alaska Task Force, General Files, 1972–1978, National Park Service, Record Group 79, National Archives and Records Administration, Anchorage, AK

NPS Records

National Park Service, Record Group 79, National Archives and Records Administration, College Park, MD

NPS Yellowstone Records

Natural and Social Sciences Records, Yellowstone National Park Heritage and Research Center, Gardiner, MT

Seidensticker Interview

John C. Seidensticker, interview by the author, Washington, DC, 25 Jan. 2007

Siniff Interview

Donald B. Siniff, interview by the author, St. Paul, MN, 16 Aug. 2004

Siniff Papers

Personal papers of Donald B. Siniff, copies in possession of the author

Smithsonian Acc. 03-037

Conservation and Research Center (National Zoological Park), Subject Files, 1970–1987, Smithsonian Accession 03-037, Smithsonian Institution Archives, Washington, DC

Smithsonian Record Unit 254	Smithsonian Institution, Assistant Secretary for Science Records, 1963–1978, Record Unit 254, Smithsonian Institution Archives, Washington, DC
Smithsonian Record Unit 271	Smithsonian Institution, Office of Environmental Sciences, Ecology Program Records, 1965–1973, Record Unit 271, Smithsonian Institution Archives, Washington, DC
Smithsonian Record Unit 329	Smithsonian Institution, Assistant Secretary for Science Records, Circa 1963–1986, Record Unit 329, Smithsonian Institution Archives, Washington, DC
Swenson and Cochran Interview	George Swenson and William W. Cochran, interview by the author, Champaign, IL, 20 Aug. 2006
Tester Interview	John R. Tester, interview by the author, St. Paul, MN, 2 Aug. 2004
Train Papers	Russell E. Train Papers, Library of Congress, Washington, DC
Warner Interview	Dwain W. Warner, interview by the author, Stanchfield, MN, 18 Aug. 2004
Warner Papers	Personal papers of Dwain W. Warner, copies in possession of the author
Wilderness Society Records	Wilderness Society Records, CONS 130, Denver Public Library, Denver, CO

Notes

INTRODUCTION: Knowing the Wild

1. Peter Matthiessen, *Wildlife in America* (New York: Viking, 1959); Gregg Mitman, *Reel Nature: America's Romance with Wildlife on Film* (Cambridge, MA: Harvard University Press, 1999); Mark V. Barrow Jr., *Nature's Ghosts: Confronting Extinction from the Age of Jefferson to the Age of Ecology* (Chicago: University of Chicago Press, 2009).

2. Gregg Mitman, "When Nature *Is* the Zoo: Vision and Power in the Art and Science of Natural History," *Osiris* 11 (1996): 117–143; Thomas R. Dunlap, *Saving America's Wildlife: Ecology and the American Mind, 1850–1990* (Princeton, NJ: Princeton University Press, 1991), on 176; John MacKenzie, *The Empire of Nature: Hunting, Conservation, and British Imperialism* (New York: Manchester University Press, 1988), on 130.

3. William Cronon, "The Trouble with Wilderness; or, Getting Back to the Wrong Nature," *Environmental History* 1 (1996): 7–28.

CHAPTER ONE: Cold War Game

1. On radiotelemetry and the Cold War–era vision of nature of American wildlife biologists, see Gregg Mitman, "When Nature *Is* the Zoo: Vision and Power in the Art and Science of Natural History," *Osiris* 11 (1996): 117–143. On Warner's career, see Kevin Winker, "In Memoriam: Dwain W. Warner, 1917–2005," *Auk* 123 (2006): 911–912; "Curriculum Vita Data," 1975, Dwain W. Warner Biographical File, University of Minnesota Archives. On Allen, see Gregg Mitman, *Reel Nature: America's Romance with Wildlife on Film* (Cambridge, MA: Harvard University Press, 1999), 122; Mark V. Barrow Jr., *A Passion for Birds: American Ornithology after Audubon* (Princeton, NJ: Princeton University Press, 1998), 189.

2. Warner Interview.

3. Dwain W. Warner to Arthur N. Wilcox, 20 Mar. 1958, Warner Papers.

4. A. N. Wilcox to Guy Stanton Ford, 9 Apr. 1940, in A. C. Hodson, *History of the Cedar Creek Natural History Area*, University of Minnesota Field Biology Program, Occasional Papers no. 2 (Minneapolis: University of Minnesota, 1985), 96–97; Theodore Blegen, "The Cedar Creek Forest Laboratory: An Address Delivered on September 14, 1957, at Cedar Creek Forest," in Hodson, *History of Cedar Creek*, 110–113, on 111. On Wilcox's long involvement with Cedar Creek, see Hodson, *History of Cedar Creek*, 26, 43.

5. On Tester's involvement, see H. G. Payne to John R. Tester, 29 May 1958, Warner Papers. On the Cedar Creek NSF grant, see A. N. Wilcox to Homer T. Mantis, Merle P. Meyer, E. F. Cook, J. R. Beer, D. B. Lawrence, J. D. Ovington, H. E. Wright Jr., and Dwain Warner, 14 Oct. 1958, Warner Papers.

6. Parker S. Trefethen, *Sonic Equipment for Tracking Individual Fish*, Special Scientific Report, Fisheries 179 (Washington, DC: FWS, 1956): 1–11.

7. William H. Marshall, "Radiotracking of Porcupines and Ruffed Grouse," in Lloyd E. Slater, ed., *Bio-Telemetry: The Use of Telemetry in Animal Behavior and Physiology in Relation to Ecological Problems* (New York: Pergamon, 1963), 173–178, on 173; "Curriculum Vitae," n.d. [1978?], William H. Marshall Biographical File, University of Minnesota Archives, Elmer L. Andersen Library, Minneapolis, MN. Aldo Leopold's encouragement of ruffed grouse research in Minnesota is described in David L. Hansen, *A Century of Research in Natural Resources* (St. Paul: University of Minnesota, Minnesota Agricultural Research Station, College of Natural Resources, 2003). See also Thomas R. Dunlap, *Saving America's Wildlife: Ecology and the American Mind, 1850–1990* (Princeton, NJ: Princeton University Press, 1991), 71–74.

8. R. W. Burwell to William H. Marshall, 27 Oct. 1958, Burwell to James W. Kimball, 23 Sept. 1958, Box 5, Marshall Papers.

9. Daniel L. Leedy to Regional Director, Minneapolis [R. W. Burwell], 30 Dec. 1958, Box 1, FWS Wildlife Research Files; Clarence Cottam to William H. Marshall, 11 Nov. 1958, C. R. Gutermuth to Marshall, 31 Mar. 1959, Box 5, Marshall Papers; Marshall to Laurence R. Lunden, 12 Jan. 1959, and attached proposal, "Development and Use of Short Wave Radio Transmitters to Trace Animal Movements," Box 5, Marshall Papers.

10. George Sprugel to Dwain W. Warner and John R. Tester, 26 Jan. 1959, Warner Papers; Sprugel to William H. Marshall, 27 Jan. 1959, Box 5, Marshall Papers. On Sprugel's Environmental Biology Program, see Toby A. Appel, *Shaping Biology: The National Science Foundation and American Biological Research, 1945–1975* (Baltimore: Johns Hopkins University Press, 2000), 22. Marshall to Sprugel, 2 Feb. 1959, Box 5, Marshall Papers. The statement about Warner's persuasiveness is William Cochran's; Cochran and Swenson Interview. Warner's recollection is from Warner Interview.

11. On Eklund's radio-thermometer research, see John H. Busser and Marion Mayer, "Radio Thermometer That Fits in a Penguin Egg," *Naval Research Reviews* (June 1957): 9–13; Carl R. Eklund and Frederick E. Charlton, "Measuring the Temperatures of Incubating Penguin Eggs," *American Scientist* 47 (1959): 80–86. On Galler's interest in biotelemetry, see S. R. Galler and C. E. Fix, "Animal Tracking Gone Modern," *Naval Research Reviews* (Aug. 1961): 11–14; Sidney R. Galler, "A Look into the Future," in Sidney R. Galler et al., eds., *Animal Orientation and Navigation* (Washington, DC: NASA, 1972), 597–601. On the Naval Research Laboratory work, see Cobert D. LeMunyan et al., "Design of a Miniature Radio Transmitter for Use in Animal Studies," *Journal of Wildlife Management* 23 (1959): 107–110.

12. The February meeting on wildlife tracking is described in Daniel L. Leedy to Drs. Buckley and Aldrich, Patuxent Refuge, 14 Jan. 1959, Box 1, FWS Wildlife Research Files; John W. Aldrich to Chief, Branch of Wildlife Research [Daniel Leedy], 19 Feb. 1959, Box 5, Marshall Papers. The quote from Galler and Hayes is in S. R. Galler and H. Hayes, "Biological Orientation," *Naval Research Reviews* (Sept. 1962): 1–4, on 3. Galler later described one of the ONR Biology Branch's main missions as "the utilization and simulation of biological phenomena for the advance of naval operations"; Sidney R. Gal-

ler, "Office of Naval Research," *AIBS Bulletin* 13 (Oct. 1963): 25–27, on 27. On ONR's changing place in the postwar federal science-funding system, see Harvey M. Sapolsky, *Science and the Navy: The History of the Office of Naval Research* (Princeton, NJ: Princeton University Press, 1990). On Midway bird strikes, see Nate Haseltine, "Gooney May Be Looney but Still Be Fun," *Washington Post and Times-Herald*, 7 Mar. 1955; William M. Blair, "U.S. Plans to Curb Periling of Planes by Midway Bird," *New York Times*, 23 Apr. 1958. In the summer of 1960, *Time* magazine reported on the potential use of radio tracking to solve the bird-strike problem: "Eventually, the Navy hopes, its little radios will signal defeat for an ancient enemy: the albatrosses (known as gooney birds) that nest by the thousands on Midway Island and make its runways dangerous for aircraft"; "Getting Rid of Gooneys," *Time*, 23 June 1960, 40.

13. On NSF's rejection of Warner and Tester's proposal, see Dwain W. Warner to John T. Wilson, 13 Mar. 1959, Warner Papers. On their successful application to the Hill Family Foundation, see Warner to Theodore C. Blegen, 2 Oct. 1959, Box 1, Cedar Creek Records; A. A. Heckman to Warner, 28 Apr. 1959, Warner Papers; Warner Interview. On Marshall's successful NSF proposal and correspondence with Seubert, see George Sprugel to William H. Marshall, 16 Mar. 1959, Sprugel to Marshall, 15 June 1959, Marshall to John L. Seubert, 6 Apr. 1959, Seubert to Marshall, 1 Apr. 1959, and Marshall to Seubert, 23 Apr. 1959, Box 5, Marshall Papers.

14. Ben East, "More Beep-Beep," *Outdoor Life* 125 (Aug. 1959): 26; William H. Marshall to D. R. Bishop, 25 Aug. 1959, Box 5, Marshall Papers.

15. On Marshall's requests to Honeywell for a site visit, see William H. Marshall to Roy H. Malm, 26 Mar. 1959, Marshall to R. J. Boyle, 25 Aug. 1959, Marshall to William Currie, 12 Nov. 1959, Box 5, Marshall Papers. On Marshall's rejection of a portable tracking system, see Marshall to Charles D. Canfield, 10 Dec. 1959, Box 5, Marshall Papers.

16. James W. Kimball to All Research Personnel, 18 Dec. 1959, and attached manuscript, James Kimball, "What the Wildlife Administrator Wants from Research," delivered 7 Dec. 1959 at the Twenty-first Midwest Wildlife Conference, Minneapolis, MN, Box 1, MN Game and Fish Research Records.

17. Eugene H. Dustman to Henry S. Mosby, cc to E. L. Cheatum, Frank Craighead, and Lowell Adams, 29 Sept. 1960, Box 2, FWS Wildlife Research Files; Lowell Adams, "History," *Wildlife Telemetry Newsletter* 1 (Jan. 1961): 1.

18. "This One's for the Birds," *St. Paul Pioneer Press*, 19 Mar. 1960; Lewis Patterson, "Wiretapping of Grouse to Bare Intimate Secrets," *St. Paul Dispatch*, 25 Mar. 1960. On the interest of magazine editors, see William H. Marshall to Milton Stenlund, 24 Mar. 1960, Box 5, Marshall Papers.

19. On Canfield's "heroics" during the installation, see William H. Marshall to James L. Morrill, 25 Apr. 1960, Box 5, Marshall Papers. On the results of the first test, see Gordon W. Gullion to Marshall, 18 May 1960, Box 5, Marshall Papers.

20. "Tiny Space-Age Radios Track Grouse," *Honeywell World* (16 May 1960): 8; Gordon W. Gullion to William H. Marshall, 20 May 1960 and 29 May 1960, Box 5, Marshall Papers.

21. Canfield's last letter to the team was to Marshall, 25 May 1960, Box 5, Marshall Papers. The tracking of "porkies" is detailed in William H. Marshall, Gordon W. Gullion, and Robert G. Schwab, "Early Summer Activities of Porcupines as Determined by Radio-Positioning Techniques," *Journal of Wildlife Management* 26 (1962): 75–79;

Gordon W. Gullion and Robert Schwab, "Plotting Animal Locations by Means of a Miniature Radio," unpublished manuscript, 7 June 1960, Gordon W. Gullion and Robert Schwab, "Further Observations on the Problems of Determining the Locations of Animals Carrying Miniature Radio Transmitters," unpublished manuscript, 9 July 1960, Gordon W. Gullion to Phillip Tichenor, 13 July 1960, Box 5, Marshall Papers.

22. Gordon W. Gullion to William H. Marshall, 12 July 1960, Phillip J. Tichenor, Press Release for Demonstration on July 20, 1960, 13 July 1960, Marshall to George Sprugel, 22 Aug. 1960, Sprugel to Marshall, 23 Aug. 1960, Box 5, Marshall Papers.

23. Warner Interview; Emil Pfender, "Ernst R. G. Eckert," in *Memorial Tributes: National Academy of Engineering*, vol. 11 (Washington, DC: National Academies Press, 2007), 109–112. The museum group report is quoted in Lowell Adams, "Where Are We?" *Wildlife Telemetry Newsletter* 1 (Jan. 1961): 3–11, on 4.

24. Eugene H. Dustman to Henry S. Mosby, cc to E. L. Cheatum, Frank Craighead, and Lowell Adams, 29 Sept. 1960, Daniel L. Leedy to Director, Patuxent Wildlife Research Center, Director, Denver Wildlife Research Center, and Leaders, Cooperative Wildlife Research Units, 8 Sept. 1960, Box 2, FWS Wildlife Research Files.

25. George W. Swenson Jr., "Looking Back: Sputnik," *IEEE Potentials* 16 (Feb.–Mar. 1997): 36–40, on 39; Cochran and Swenson Interview; Rexford D. Lord Jr., "Miniature Radio Tracking System," *Wildlife Telemetry Newsletter* 1 (July 1961): 3–6; Rexford D. Lord Jr., "W-42-R-9," *Monthly Wildlife Research Letter, Department of Conservation and [Illinois] Natural History Survey, Cooperating* 3 (Mar. 1960): 1.

26. Cochran and Swenson Interview; Swenson, "Looking Back: Sputnik," 36–40, on 39; Rexford D. Lord Jr., "W-42-R-10," *Monthly Wildlife Research Letter, Department of Conservation and [Illinois] Natural History Survey, Cooperating* 4 (Feb. 1961): 1.

27. Adams, "History," 1–2, on 2; Lowell Adams, "Special Meeting on Radio Tracking of Wildlife," *Wildlife Telemetry Newsletter* 1 (July 1961): 1–3, on 2; Lowell Adams, "What Are the Problems?" *Wildlife Telemetry Newsletter* 1 (Jan. 1961): 10–12, on 10; Rexford D. Lord Jr., "Radio-Tracking a Flying Duck," undated, unpaginated manuscript, copy in possession of the author, received 26 Feb. 2007.

28. The quote from Lord about soil attenuation is in Lord, "Radio-Tracking a Flying Duck." On the performance of Cochran's system, see Lord, "Miniature Radio Tracking System," 3–7; William W. Cochran and Rexford D. Lord Jr., "A Radio-Tracking System for Wild Animals," *Journal of Wildlife Management* 27 (1963): 9–24. On the mammalogy meeting, see R. D. Lord Jr. and D. A. Casteel, "W-42-R-10," *Monthly Wildlife Research Letter, Department of Conservation and [Illinois] Natural History Survey, Cooperating* 4 (June 1961): 1–2. On the number of letters received, see R. D. Lord Jr., and D. A. Casteel, "Rabbit Management," *Monthly Wildlife Research Letter, Department of Conservation and [Illinois] Natural History Survey, Cooperating* 4 (Sept. 1961): 2. Cochran's recollection is from Cochran and Swenson Interview.

29. Ralph J. Ellis, "Investigations of Furbearers," *Monthly Wildlife Research Letter, Department of Conservation and [Illinois] Natural History Survey, Cooperating* 4 (Sept. 1961): 2–3; Rexford D. Lord Jr., Frank C. Bellrose, and William W. Cochran, "Radiotelemetry of the Respiration of a Flying Duck," *Science* 137 (1962): 39–40; Lord, "Radio-Tracking a Flying Duck"; Cochran and Swenson Interview. Cochran's description is in discussion following Klaus Schmidt-Koening, "The Problem of Distant Tracking in Experiments in Bird Orientation," in Slater, *Bio-Telemetry*, 119–124, on 124.

30. "Grousar Radio Project: A Progress Report, July 26, 1961–November 22, 1961,"

Box 5, Marshall Papers; William H. Marshall, "Radiotracking of Porcupines and Ruffed Grouse," in Slater, *Bio-Telemetry*, 173–178, on 177; Cochran Interview. On tensions at Carlos Avery, see Arnold B. Erickson to John B. Moyle, 18 Dec. 1961, Box 2, MN Game and Fish Research Records; Erickson to Moyle, 1 Feb. 1962, Box 3, MN Game and Fish Research Records.

31. Cochran Interview; Tester Interview. The Sandersons' work is described in Glen C. Sanderson and Beverly C. Sanderson, "Radio Tracking Rats in Malaya: A Preliminary Study," *Journal of Wildlife Management* 28 (1964): 752–768; Laurie Sanderson to Etienne Benson, e-mail, 17 June 2007, in possession of the author. After Cochran returned to the Illinois Natural History Survey in May 1964, Swenson recalled, he "had biologists from all over the country coming in. . . . He built a room in his basement where visitors could come and stay while they learned his technology, and instead of instead of charging them, he would just teach them. Then they'd go away and build transmitters, and we all thought that was pretty generous"; Cochran and Swenson interview.

32. Lowell Adams, "Telemetry Meetings," *Wildlife Telemetry Newsletter* 1 (Aug. 1962): 1–7; "Progress Report: Development and Use of Short Wave Radio Transmitters to Trace Animal Movements, August 1960–February 1962," 1 Mar. 1962, Box 5, Marshall Papers.

33. Jack Wilson, "Secrets of the Birds May Guide Missiles," *Minneapolis Tribune*, 7 Sept. 1961; "Undersea Lag Is Noted," *New York Times*, 9 Nov. 1961; Cochran and Swenson Interview.

34. The conference was announced in Adams, "Where Are We?" 3–11, on 8; Lowell Adams, "Radio Tracker Developments," *Wildlife Telemetry Newsletter* 1 (Jan. 1962): 2–4. D. W. Warner, "Fundamental Problems in the Use of Telemetry in Ecological Studies," in Slater, *Bio-Telemetry*, 11–24, on 17, also cited in Mitman, "When Nature *Is* the Zoo," 139. Warner, "Fundamental Problems," 20. Schmitt's comment is in discussion following Robert R. Jeffries, "The Frontiers of Telemetry Technology," in Slater, *Bio-Telemetry*, 65–74, on 73. Galler's comment is in discussion following Warner, "Fundamental Problems," on 24.

35. Cochran Interview; Warner Interview; Dwain W. Warner, "Preliminary Request for Research Grant and for Instrument Space on a Satellite in Polar Orbit," Grant Proposal to NASA, 16 Apr. 1962, Warner Papers.

36. On the continuing tensions at Carlos Avery, see John B. Moyle to Dwain W. Warner, 4 June 1962, Moyle to James W. Kimball, 15 June 1962, Box 3, MN Game and Fish Research Records. On Marshall's directorship, see "'U' of Minnesota Promotes 171 Faculty Members," Press Release, University of Minnesota News Service, 25 June 1962, William H. Marshall Biographical File, University of Minnesota Archives. On the antenna towers, see Warner to Bryce Crawford Jr., 30 July 1962, Box 1, Cedar Creek Records.

37. William H. Marshall, "Notes on Field Trip with Dr. Warner and Bill Cochrane [*sic*] to Cedar Creek (Afternoon of August 20) to Discuss the Activation of Research Project #62–4," n.d. [1962], Marshall to William W. Cochran, 5 Nov. 1962, Box 1, Cedar Creek Records.

38. Warner Interview; Dwain W. Warner, "Space Tracks: Bioelectronics Extends Its Frontiers," *Natural History* 62 (1963): 8–15; Charles A. Kemper, "The Minnesota Ornithologists Union Annual Winter Meeting: A Report," *Passenger Pigeon* 24 (1962): 91–92, on 91; Dennis G. Raveling to Dwain W. Warner, 1 Oct. 1986, Warner Papers. Raveling's

letter was a contribution to a scrapbook of memorabilia collected for presentation at a retirement dinner held in Warner's honor at the University of Minnesota.

39. Bryce Crawford Jr., to Marshall, 16 Nov. 1962, Marshall to Dwain Warner, 18 Feb. 1963, Box 1, Cedar Creek Records; Warner to Marshall, 11 Mar. 1963, Cedar Creek Files; Warner Interview.

40. Olin L. Kaupanger to James W. Kimball, 23 July 1962, William H. Marshall to Daryle M. Feldmeir, 26 Mar. 1963, Feldmeir to Marshall, 27 Mar. 1963, Box 5, Marshall Papers; Jim Kimball, "Electronics Permits Intimate Insights into Animal Life," *Minneapolis Star Tribune*, 22 Mar. 1963.

41. Lowell Adams, "Notice!" *Wildlife Telemetry Newsletter* 2 (Feb. 1963): 1; Lowell Adams, "Wildlife Telemetry Meeting," *Wildlife Telemetry Newsletter* 2 (May 1963): 1–4; Tester Interview.

42. Tester Interview; Warner Interview; "Cedar Creek Revisited," *Northwest Area Foundation Newsletter* 5 (Spring 1982): 2–3. In 1967, the Cedar Creek telemetry group received an NIH training grant at a level of $600,000 over five years; Donald B. Siniff to Lee L. Eberhardt, 3 Aug. 1967, Siniff Papers.

43. Lowell Adams, "An Editorial," *Wildlife Telemetry Newsletter* 3 (Dec. 1963): 6–7, on 6; Lowell Adams to John Olive, 30 Dec. 1963, Box 3, Marshall Papers.

44. "Progress Report: Studies of Movements, Behavior and Activities of Ruffed Grouse Using Radio Telemetry Techniques, 1963," on 20, Box 5, Marshall Papers.

45. Gordon W. Gullion to William H. Marshall, 9 Jan. 1964, Box 3, Marshall Papers; "Progress Report: Studies of Movements, Behavior and Activities of Ruffed Grouse Using Radio Telemetry Techniques, 1964," Box 5, Marshall Papers; Geoffrey A. Godfrey and William H. Marshall, "Brood Break-Up and Dispersal of Ruffed Grouse," *Journal of Wildlife Management* 33 (1969): 609–620, on 610; William H. Marshall, "Ruffed Grouse Behavior," *BioScience* 15 (Feb. 1965): 92–94. Marshall reported the most recent results of the Grousar project at the March 1964 telemetry session of the North American Wildlife Conference; after a peak attendance of about three hundred in 1963, attendance had dropped off to less than seventy people, only fourteen of whom were actively engaged in radio-tracking projects. Lowell Adams, "Wildlife Telemetry Meeting," *Wildlife Telemetry Newsletter* 3 (Apr. 1964): 1–2.

46. William W. Cochran, Dwain W. Warner, and John R. Tester, "The Cedar Creek Automatic Radio Tracking System," Minnesota Museum of Natural History Technical Report no. 7, May 1964, 9 pp., on p. 5; Warner to William H. Marshall, 14 Jan. 1964, Cedar Creek File; Gordon W. Gullion to Marshall, 29 Oct. 1963, Box 3, Marshall Papers.

47. Tester Interview; Mech Interview; Cochran Interview; William W. Cochran, Dwain W. Warner, and John R. Tester, "The Cedar Creek Automatic Radio Tracking System," Minnesota Museum of Natural History Technical Report no. 7, May 1964, 9 pp., on p. 3; L. David Mech et al., "A Collar for Attaching Radio Transmitters to Rabbits, Hares, and Raccoons," *Journal of Wildlife Management* 29 (1965): 898–902. The quote on "exasperatingly clever" raccoons is in A. Karl Slagle, "Designing Systems for the Field," *BioScience* 15 (Feb. 1965): 109–111, on 109.

48. Warner Interview; Cochran Interview; Tester Interview. For an example of Graber's concern about tag weight, see Richard R. Graber and Steven L. Wunderle, "Telemetric Observations of a Robin," *Auk* 83 (1966): 674–677. On Marshall's big-game hunt-

ing in New Zealand, see William H. Marshall to Bob [Robert W. Seabloom], 10 Dec. 1960, Box 5, Marshall Papers.

49. Gordon W. Gullion to William H. Marshall, 29 Oct. 1963, Box 3, Marshall Papers; Kuechle Interview; Alan B. Sargeant, James E. Forbes, and Dwain W. Warner, "Accuracy of Data Obtained through the Cedar Creek Automatic Radio Tracking System," Minnesota Museum of Natural History Technical Report no. 10, Dec. 1965, 20 pp., on p. 7. On power lines and the relative lack of signal distortion, see Cochran, Warner, and Tester, "Cedar Creek Automatic Radio Tracking System," 6; on wind, see William W. Cochran et al., "Automatic Radio-Tracking System for Monitoring Animal Movements," *BioScience* 15 (Feb. 1965): 98–99.

50. Cochran and Swenson Interview; Cochran Interview. On Cochran's work after returning to Illinois, see Lowell Adams, "Letters to the Editor," *Wildlife Telemetry Newsletter* 3 (Apr. 1964): 4–5, on 5; R. E. Hawkins and G. G. Montgomery, "'Movements of Translocated Deer as Determined by Telemetry," *Journal of Wildlife Management* 33 (1969): 196–203; W. W. Cochran, G. G. Montgomery, and R. R. Graber, "Migratory Flights of *Hylocichla* Thrushes in Spring: A Radiotelemetry Study," *Living Bird* 6 (1967): 213–225, on 224.

51. Cochran, Warner, and Tester, "Cedar Creek Automatic Radio Tracking System," 1; J. R. Tester, J. W. Ternes, and D. B. Siniff, "Behavior and Mortality of Free-Ranging Raccoons, Snowshoe Hares, and Striped Skunks after Exposure to 300 R γ Radiation," *Radiation Research* 69 (1977): 500–512.

52. On the film-reading process and "girls," see Hodson, *History of Cedar Creek*, 91; Donald B. Siniff to John R. Tester, 27 Jan. 1970, Siniff Papers; Siniff Interview. For estimates of data processing time, see Dwain W. Warner to William H. Marshall, 14 Jan. 1964, Cedar Creek Files; Cochran, Warner, and Tester, "Cedar Creek Automatic Radio Tracking System," 8.

53. Siniff Interview; Lee L. Eberhardt to Donald B. Siniff, 5 May 1964, Siniff Papers; L. David Mech and John R. Tester, "Biological, Behavioral, and Physical Factors Affecting Home Ranges of Snowshoe Hares (*Lepus americanus*), Raccoons (*Procyon lotor*), and White-Tailed Deer (*Odocoileus virginianus*) under Natural Conditions, Progress Report No. 1," Museum of Natural History Technical Report no. 9, July 1965; Alan B. Sargeant to Siniff, 10 Oct. 1964, Siniff Papers; Donald B. Siniff, "Computer Programs for Analyzing Radio Tracking Data," Minnesota Museum of Natural History Technical Report no. 12, Mar. 1966.

54. Donald B. Siniff, "A Simulation Model of Animal Movement Data," James Ford Bell Museum of Natural History Technical Report no. 14, July 1967. In March 1965, Tester and Siniff sent a copy of a paper they had presented at the North American Wildlife Conference to Hayne and Dice, asking for their comments and inviting them to visit Cedar Creek; Donald B. Siniff to Lee L. Eberhardt, 3 Aug. 1967, Siniff to Don Hayne, 30 Mar. 1965, Hayne to Siniff, 24 May 1965, Lee R. Dice to Siniff, 10 Apr. 1965, Siniff Papers.

55. Glen C. Sanderson, "The Study of Mammal Movements: A Review," *Journal of Wildlife Management* 30 (1966): 215–235, on 221, 223.

56. J. R. Tester, D. B. Siniff, and C. R. Jessen, "Use of Telemetry as a Means of Studying Spacing and Behavior of Animals," in Aristide H. Esser, ed., *Behavior and Environment: The Use of Space by Animals and Men* (New York: Plenum, 1971), 117–119, on 118; Siniff Interview; Cochran Interview.

57. Lee L. Eberhardt to Donald B. Siniff, 16 Mar. 1965, Siniff to Eberhardt, 21 Dec. 1966, Siniff Papers; Siniff Interview. Siniff remained interested in simulation for several years after completing his thesis. In 1970, for example, he attended the two-week workshop on "Simulation Games for Teaching of Natural Resources Management" at the University of Washington. However, his focus soon shifted elsewhere; see D. B. Siniff and C. Jessen, "A Simulation Model of Animal Movement Patterns," *Advances in Ecological Research* 6 (1969): 185–219; Donald B. Siniff, "Computer Simulation of Animal Populations," Application for Grant in Aid of Research to the Graduate School, University of Minnesota, 15 Feb. 1971, 4 pp., on p. 2, Siniff Papers.

58. Warner Interview; William W. Cochran, "No Question—No Answer Corner," *Wildlife Telemetry Newsletter* 4 (Jan. 1965): 5.

59. Josephine K. Doherty to William H. Marshall, 7 Oct. 1964; John S. Rankin Jr. to Marshall, 18 Jan. 1965, "Proposal Evaluation Sheet," n.d. [Spring 1965], Marshall to Josephine K. Doherty, 14 Apr. 1965, Harve J. Carlson to Marshall, 1 Apr. 1965, William H. Marshall, "Progress Report: Studies of Movements, Behavior and Activities of Ruffed Grouse Using Radio Telemetry Techniques, 1965," Box 5, Marshall Papers.

60. Gordon W. Gullion to William H. Marshall, 10 Sept. 1965, Box 5, Marshall Papers; Gullion to Marshall, 4 Oct. 1965, Box 3, Marshall Papers.

61. William H. Marshall to John Kupa, Robert Brander, Geoffrey Godfrey, Ronald Barrett, and Philip Schladweiler, 5 Apr. 1966, William H. Marshall, "Final Report—Development and Use of Radio Telemetry Techniques to Study Ruffed Grouse," n.d. [Spring 1967], William H. Marshall, "Progress Report: Studies of Movements, Behavior and Activities of Ruffed Grouse Using Radio Telemetry Techniques, 1965," Sidney L. Markusen to Marshall, 10 Oct. 1965, Box 5, Marshall Papers.

62. William H. Marshall to Donald B. Lawrence, James R. Beer, and Francis Spurrell, 23 Nov. 1965, with attached John R. Tester, "Application for Research Use of the Cedar Creek Natural History Area, 62–4, Supplement," 3 Nov. 1965, 3 pp., Cedar Creek Files; the quote on "sterility or death" is on p. 1, the comparison to natural causes, p. 3.

63. William H. Marshall to Donald B. Lawrence, James R. Beer, and Francis Spurrell, 23 Nov. 1965, Cedar Creek Files. On Lawrence's long history with Cedar Creek and complex position on experimentation, see Hodson, *History of Cedar Creek*, 5, 35, 80–81, 99; Donald B. Lawrence, "Historical Perspectives, Objectives, and Obligations," in Hodson, *History of Cedar Creek*, 7–10, on 9.

64. Donald Lawrence to William H. Marshall, 27 Nov. 1965, Marshall to Lawrence, 30 Nov. 1965, Cedar Creek Files.

65. John R. Tester to William Marshall, 25 Jan. 1966, Marshall to Don Lawrence, Dave Merrell, Bill Reiners, James Beer, and Francis Spurrell, cc to Tester, 31 Jan. 1966, Lawrence to Marshall, 23 Feb. 1966, Cedar Creek Files.

66. Minutes of Cedar Creek Advisory Board, 3 Mar. 1966, Box 1, Cedar Creek Records.

67. Keith L. Heezen and John R. Tester, "Evaluation of Radio-Tracking by Triangulation with Special Reference to Deer Movements," *Journal of Wildlife Management* 31 (1967): 124–141, on 124, 140; Mech Interview. Robert Kenward, an ornithologist who played an important role in introducing radiotelemetry to the United Kingdom in the 1970s, later recalled that the growing awareness of systematic biases inherent in automatic systems with fixed antennas limited the amount of important scientific research that could be done with such systems; Kenward Interview.

68. William H. Marshall, "Commercial Suppliers," *Wildlife Telemetry Newsletter* 5 (Dec. 1966): 2–4, on 3; William H. Marshall, "Closing Editorial," *Wildlife Telemetry Newsletter* 5 (Dec. 1966): 8; William H. Marshall, "Committee Business," *Wildlife Telemetry Newsletter* 5 (Sept. 1967): 1–2, on 1; William H. Marshall, "Committee Meeting," *Wildlife Telemetry Newsletter* 5 (Sept. 1967): 2–4.

69. Mech Interview; L. David Mech and L. D. Frenzel, "The Mechanics and Significance of Timber Wolf–Deer Coactions," proposal to NSF, 15 Dec. 1966, with attached review by Marshall, 31 Jan. 1967, Box 8, Marshall Papers. For early results of Mech's wolf research, see L. David Mech, *The Wolf: The Ecology and Behavior of an Endangered Species* (Garden City, NY: Published for the American Museum of Natural History by Natural History Press, 1970); L. David Mech, "Mortality, and Population Trends of Wolves in Northeastern Minnesota," *Journal of Mammalogy* 58 (1977): 559–574. Biologists in Ontario had begun radio-tracking wolves a few years before Mech began his work in Minnesota; George B. Kolenosky and David H. Johnston, "Radio-Tracking Timber Wolves in Ontario," *American Zoologist* 7 (1967): 289–303. Siniff Interview; Donald B. Siniff to Lee Eberhardt, 3 Aug. 1967, Siniff to William H. Marshall, 5 Feb. 1970, Siniff Papers.

70. William H. Marshall, "The Cedar Creek Natural History Area: A Progress Report," *Journal of the Minnesota Academy of Science* 35 (1968): 57–60, on 58; Tester, Ternes, and Siniff, "Behavior and Mortality," 501–502; Richard Huempfner to William H. Marshall, 5 Mar. 1970, Siniff Papers.

71. On the building of radio-tracking equipment for other groups, see Tester Interview. On international visitors, see William H. Marshall to Donald B. Siniff, 18 Sept. 1968, Siniff Papers; Vladimir Sokolov to Dwain W. Warner, 31 Jan. 1968, Warner Papers; Warner to Etienne Benson, e-mail, 30 Oct. 2004, in possession of the author; Marshall to Robert Jordan, Frank McKinney, and John Tester, 2 Aug. 1971, Box 8, Marshall Papers. For Tester's comment on software, see Tester to J. J. Lynch, 7 Dec. 1970, Box 8, Marshall Papers.

72. On Tester's activity pattern work, see J. R. Tester, "Analysis of Circadian Rhythms of Freeranging Mammals," in H.-J. Klewe and H. P. Kimmich, eds., *Biotelemetry IV: Proceedings of the Fourth International Symposium on Biotelemetry, Garmisch-Partenkirchen, Germany, May 28–June 2, 1978* (Braunschweig: Döring, 1978): 167–170. Hodson claims that 1977 "marked the beginning of extensive uses of telemetry equipment developed at Cedar Creek for radio-telemetry research in other parts of the country mostly by students of the principal investigators," but such work had begun significantly earlier; Hodson, *History of Cedar Creek*, 66. On Sunquist's work, see Melvin E. Sunquist, *The Social Organization of Tigers (Panthera Tigris) in Royal Chitawan National Park, Nepal*, Smithsonian Contributions to Zoology 336 (Washington, DC: Smithsonian Institution Press, 1981); Fiona Sunquist and Melvin E. Sunquist, *Tiger Moon* (Chicago: University of Chicago Press, 1988). On the importance of the Bioelectronics Laboratory to the Nepal project, see Donald B. Siniff to Ross Simons, 14 Apr. 1976, Siniff Papers. On the sea otter research, see David L. Garshelis and Donald B. Siniff, "Evaluation of Radio-Transmitter Attachments for Sea Otters," *Wildlife Society Bulletin* 11 (1983): 378–383; "Cedar Creek Revisited," *Northwest Area Foundation Newsletter* 5 (Spring 1982): 2–3, on 3. The Northwest Area Foundation was the new name of the Hill Family Foundation, which had supported the Cedar Creek group in the early 1960s.

73. Kuechle Interview; Siniff Interview.

74. Tester Interview; Siniff Interview; Kuechle Interview; Cochran Interview. On efforts to update the Cedar Creek system, see Donald B. Lawrence, "Historical Perspectives, Objectives, and Obligations," in Hodson, *History of Cedar Creek*, 7–10, on 10. The founding of Wildlife Materials is briefly described at wildlifematerials.com. On the Denver laboratory, see Olin E. Bray, Richard E. Johnson, and A. Lawrence Kolz, "A Removable Car-Top Antenna System for Radio-Tracking Birds," *Bird-Banding* 46 (1971): 15–18. V. B. Kuechle, "State of the Art of Biotelemetry in North America," in C. L. Cheeseman and R. B. Mitson, eds., *Telemetric Studies of Vertebrates: The Proceedings of a Symposium Held at the Zoological Society of London on 21 and 22 November, 1980* (New York: Academic, 1982): 1–18, on 1, 4.

75. Kenward Interview; A. N. Lance and A. Watson, "A Comment on the Use of Radio Tracking in Ecological Research," in Charles J. Amlaner Jr. and David W. Macdonald, eds., *A Handbook on Biotelemetry and Radio Tracking* (Oxford: Pergamon, 1980), 355–359, on 355; Alan B. Sargeant, "Approaches, Field Considerations and Problems Associated with Radio Tracking Carnivores," in Amlaner and Macdonald, *Handbook on Biotelemetry*, 57–63, on 58.

76. Kuechle Interview; Hodson, *History of Cedar Creek*, on 77.

77. Tester Interview; Siniff Interview; David F. Parmelee, "A Progress Report of the Cedar Creek Natural History Area," *Journal of the Minnesota Academy of Science* 49 (1983–1984): 1–17.

78. Tester Interview; *Role for Research* (Washington, DC: FWS, 1982), Identifier 22.83, Moving Images Relating to Wildlife Issues, Compiled 1951–1992, Record Group 22, Records of the FWS, National Archives and Records Administration, College Park, MD; James J. Kennedy, "Viewing Wildlife Managers as a Unique Professional Culture," *Wildlife Society Bulletin* 13 (Winter 1985): 571–579, on 573; Tester Interview.

CHAPTER TWO: The Poetry of Wilderness

1. A. W. Erickson, "Techniques for Live-Trapping and Handling Black Bears," *Transactions of the North American Wildlife Conference* 22 (1957): 520–543, cited in John J. Craighead et al., "Trapping, Immobilizing and Color-Marking Grizzly Bears," *Transactions of the North American Wildlife Conference* 25 (1960): 347–363, on 350.

2. For Craighead's discussions with Park Service staff about the grizzly bear study, see Edmund B. Rogers to John J. Craighead, 30 June 1954, Craighead to Gordon Fredine, 7 Jan. 1958, David de L. Condon to Craighead, 23 Oct. 1958, Chief Park Naturalist [David de L. Condon] to Superintendent, Yellowstone National Park, 8 Dec. 1958, Box N-91, NPS Yellowstone Records; Craighead to Fredine, 1 Oct. 1958, Box N-371, NPS Yellowstone Records; James A. Pritchard, *Preserving Yellowstone's Natural Conditions: Science and the Perception of Nature* (Lincoln: University of Nebraska Press, 1999), 193–194. For the quote on "artificiality," see Craighead et al., "Trapping, Immobilizing and Color-Marking Grizzly Bears," on 348.

3. On the Craigheads' techniques for trapping and immobilizing bears, see Frank C. Craighead Jr., *Track of the Grizzly* (San Francisco: Sierra Club Books, 1979), 8; Pritchard, *Preserving Yellowstone's Natural Conditions*, 238.

4. Seidensticker Interview; Frank Craighead and John Craighead, "Knocking Out Grizzly Bears for Their Own Good," *National Geographic* 118 (1960): 276–291, on 287.

5. On the visibility of the tags, see Merrill D. Beal to John Craighead, 23 July 1959,

Superintendent, Yellowstone National Park [Lemuel A. Garrison], to All Park Personnel, 9 Aug. 1959, Box N-91, NPS Yellowstone Records; Memo for the Press from Yellowstone National Park, 21 Aug. 1959, Box N-371, NPS Yellowstone Records; Craighead and Craighead, "Knocking Out Grizzly Bears," on 289. On the number of bears tagged or killed, see Craighead and Craighead, "Knocking Out Grizzly Bears," 277; Craighead et al., "Trapping, Immobilizing and Color-Marking Grizzly Bears," 358–359; Craighead to Lemuel A. Garrison, 9 July 1959, Box N-91, NPS Yellowstone Records.

6. Craighead and Craighead, "Knocking Out Grizzly Bears," 278–279, 291.

7. Craighead, *Track of the Grizzly,* 10; Lowell Adams, "Where Are We?" *Wildlife Telemetry Newsletter* 1 (Jan. 1961): 3–11.

8. Robert M. McIntyre to John J. Craighead, 16 Mar. 1960, Regional Chief of Interpretation to Superintendent, Yellowstone National Park, 1 Apr. 1960, Box N-371, NPS Yellowstone Records; Chief Park Naturalist [Robert M. McIntyre] to John J. Craighead, 17 May 1960, Box N-91, NPS Yellowstone Records.

9. Craighead, *Track of the Grizzly,* 8; Superintendent, Yellowstone National Park, to Director, National Park Service, 13 Oct. 1961, Yellowstone National Park Monthly Report for October 1960, 6 Dec. 1960, Box 319, NPS Records; Adams, "Where Are We?" 9. On delays because of the need for Federal Communications Commission permits, see P. E. Smith to Superintendent, Yellowstone National Park, 19 Aug. 1960, Smith to National Park Service Chief of Design and Construction, 23 Aug. 1960, Acting Chief Engineer to Chief, Western Office, Division of Design and Construction, National Park Service, San Francisco, CA, 31 Aug. 1960, Smith to Superintendent, Yellowstone National Park, 12 Oct. 1960, Box N-371, NPS Yellowstone Records.

10. David C. Nutt to Adolph J. Murie, 1 Dec. 1960, Frederick C. Dean to the Arctic Institute of North America, Re: "Progress Report on 'Investigations on Grizzly Bear in Interior and Arctic Alaska,'" 26 Oct. 1957, attached to Dean to Murie, 4 Nov. 1957, Adolph Murie, "Comments by Adolph Murie on the Proposed Research 'Investigations of Grizzly Bear in Interior and Arctic Alaska' by Dr. Frederick C. Dean," n.d. [1960–1961], Box 6, Murie (Adolph) Files.

11. Adolph Murie, "Comments by Adolph Murie on the Proposed Research 'Investigations of Grizzly Bear in Interior and Arctic Alaska' by Dr. Frederick C. Dean"; Olaus J. Murie, "Statement Regarding Marking of Bears in Mt. McKinley National Park for Scientific Purposes," n.d. [1960–1961], Box 3, Murie (Adolph) Files.

12. Robert W. Mason to Adolph Murie, 15 Dec. 1960, Mason to Murie, 24 Jan. 1961, Lawrence C. Merriam to Olaus J. Murie, 6 Jan. 1961, Box 6, Murie (Adolph) Files.

13. Superintendent, Yellowstone National Park, to Coleman C. Newman, 31 July 1961, Memo for the Press from Yellowstone National Park, 8 Aug. 1961, Memo for the Press from Yellowstone National Park, 20 Oct. 1961, Box N-371, NPS Yellowstone Records; Yellowstone National Park Monthly Report for August, 15 Sept. 1961, Yellowstone National Park Monthly Report for September, 17 Oct. 1961, Box 104, NPS Records; Yellowstone National Park Monthly Report for October, 24 Nov. 1961, Box 319, NPS Records; Frank C. Craighead, John J. Craighead, and Richard S. Davies, "Radiotracking of Grizzly Bears," in Lloyd E. Slater, ed., *Bio-Telemetry: The Use of Telemetry in Animal Behavior and Physiology in Relation to Ecological Problems* (New York: Pergamon, 1963), 132–148, on 137; Craighead, *Track of the Grizzly,* 10, 16–18.

14. The various tracking methods are described in Frank C. Craighead Jr. and John J. Craighead, "Tracking Grizzly Bears," *BioScience* 15 (Feb. 1965): 88–92, on 91. For the

quotes from Frank Craighead and nicknames of bears, see Craighead, *Track of the Grizzly*, 26–27, 16–17, 19–20, 27.

15. Craighead, *Track of the Grizzly*, 14; Craighead, Craighead, and Davies, "Radio-tracking of Grizzly Bears," 146.

16. Olaus J. Murie, "Wilderness Philosophy, Science, and the Arctic National Wildlife Range," n.d. [1961], unpublished manuscript presented at the 12th Alaskan Science Conference, Fairbanks, AK, 28 Aug. 1–Sept. 1961, Olaus J. Murie to Howard Zahniser, 16 Sept. 1961, Box 1, Wilderness Society Records.

17. Olaus Murie's December 1961 letter is mentioned in Lawrence C. Merriam to Olaus J. Murie, 24 Jan. 1962, Box 6, Murie (Adolph) Files. On Troyer's work on Kodiak, see Will Troyer, *Into Brown Bear Country* (Fairbanks: University of Alaska Press, 2005), xii–xiii; see also Will Troyer, *Bear Wrangler: Memoirs of an Alaska Pioneer Biologist* (Fairbanks: University of Alaska Press, 2008).

18. Dean Interview; Olaus J. Murie to Conrad L. Wirth, 4 Apr. 1962, Box 1, Wilderness Society Records; Acting Regional Chief of Interpretation [Richard G. Prasil] to Regional Field Biologist [Adolph Murie], Re: "Toklat Grizzly Bear Research," 26 Feb. 1962, Box 6, Murie (Adolph) Files; Regional Field Biologist [Adolph Murie] to Regional Chief of Interpretation, Re: "Toklat Grizzly Bear Research," 1 Mar. 1962, Regional Field Biologist [Adolph Murie] to Regional Chief of Interpretation, Re: "Comments on Tagging of Grizzlies in National Parks," 5 Mar. 1962, Box 2, Murie (Adolph) Files. See Pritchard, *Preserving Yellowstone's Natural Conditions*, 241.

19. Regional Field Biologist [Adolph Murie] to Regional Director, Region IV, Re: "Comments on Biological Research Program," 5 Mar. 1962, Box 3, Murie (Adolph) Files. In January 1962, Adolph Murie told the regional chief of interpretation that "in addition to the gathering of annual statistics and observations in relation to the stated objective of the original field studies, I have always felt that there was something else for which to strive. In my work I have tried to be especially alert for observations that would create interest in the park idea and humanize man's attitude toward wildlife. Only by prolonged field studies can one accumulate such material. Here the esthetics are involved, and the 'reverence of life.' Intimate presentations are difficult to secure in quantity. One must be satisfied with quality"; Adolph Murie to Regional Chief of Interpretation, 18 Jan. 1962, Box 3, Murie (Adolph) Files. See also Pritchard, *Preserving Yellowstone's Natural Conditions*, 206–207. Olaus Murie expressed similar ideas to the director of the Park Service later that spring; Olaus J. Murie to Conrad L. Wirth, 4 Apr. 1962, Box 1, Wilderness Society Records.

20. Sellars, *Preserving Nature in the National Parks: A History* (New Haven, CT: Yale University Press, 1997), on 200, 214; Pritchard, *Preserving Yellowstone's Natural Conditions*, 202–207, 211.

21. Olaus J. Murie to David R. Brower, 18 June 1960, Carton 92, Sierra Club Records, 71/103c, Bancroft Library, University of California, Berkeley; Adolph Murie to Anthony Wayne Smith, 27 Jan. 1963, Box 21, Murie (Adolph) Files.

22. Olaus J. Murie to Angus Cameron, 24 Apr. 1963, Box 5, Murie (Olaus) Papers. See also Gregg Mitman, "When Nature *Is* the Zoo: Vision and Power in the Art and Science of Natural History," *Osiris* 11 (1996): 125. The phrase "dynamic biological complexes" is in National Research Council, Advisory Committee to the National Park Service on Research, *A Report by the Advisory Committee to the National Park Service on Research* (Washington, DC: National Academy of Science, 1963), on 21; also quoted in Sellars,

Preserving Nature, 215. One of the members of the committee was Stanley A. Cain, who told Adolph Murie in April 1963 that he agreed that the "organization man, the team man, undoubtedly gets industrial research done, and I can envision a crew working on inventory and mapping in the Parks, but for most jobs I'd like to put my blue chips on the free and unfettered student of nature, qualified but following his own inclinations and methods to a large extent"; Stanley A. Cain to Adolph Murie, 28 Apr. 1963, Box 4, Murie (Adolph) Files. The committee's skepticism toward the role of wildlife management in the parks was evident on June 14, 1963, when it heard John Craighead and graduate student Hornocker present the most recent developments on their grizzly research during a tour of Yellowstone and Grand Teton. Craighead told the committee that the radio-tracking system had been "perfected" the year before, but the committee was less interested in the technical details than in the policy implications of the Craigheads' research. One committee member asked whether it was the Craigheads' "main business to produce more grizzly bears for hunters? . . . Would it be more difficult for you to do it somewhere else?" Minutes of the National Academy of Sciences National Park Service Research Committee, Yellowstone–Grand Teton National Parks, 13–16 June 1963, Box 25, Hubbs Papers.

23. Sellars, *Preserving Nature*, 215–217. Murie's criticisms of the Leopold Report were reported in "Leopold Report Appraised," *Living Wilderness* 28 (Spring–Summer 1963): 20–22, on 21. Wilderness as a "fancy" is in "Guardians Not Gardeners," *Living Wilderness* 28 (Spring–Summer 1963): 2.

24. Adolph Murie to Richard Prasil, 6 Nov. 1963, Box 2, Murie (Adolph) Files; Adolph Murie, "A Plea for Idealism in the National Parks: A Critique," 14 Jan. 1964, Box 21, Murie (Adolph) Files. On Leopold's speech, see Pritchard, *Preserving Yellowstone's Natural Conditions*, 213–215. Howard Zahniser to Adolph Murie, 27 Jan. 1964, Box 25, Murie (Adolph) Files. See also Mark W. T. Harvey, *Wilderness Forever: Howard Zahniser and the Path to the Wilderness Act* (Seattle: University of Washington Press, 2005), 252; David J. Parsons, "The Challenge of Scientific Activities in Wilderness," in Stephen F. McCool et al., eds., *Proceedings of the Wilderness Science in a Time of Change Conference*, vol. 3, *Wilderness as a Place for Scientific Inquiry, 23–27 May 1999* (Ogden, UT: U.S. Department of Agriculture, Forest Service, Rocky Mountain Research Station, 2000), 252–257, on 253.

25. John J. Craighead, Frank C. Craighead, and Maurice G. Hornocker, "An Ecological Study of the Grizzly Bear: Fifth Annual Report, Summary of Work Accomplished, 1963," Montana Cooperative Wildlife Research Unit, Montana State University, Missoula; Acting Superintendent, Yellowstone [Luis A. Gastellum] to Regional Director, Midwest Region, Re: "An Ecological Study of the Grizzly Bear," n.d., Acting Assistant Regional Director, Midwest Region, to Director, National Park Service, cc to Superintendent, Yellowstone, Re: "Grizzly Bear Ecology, Yellowstone," 28 Feb. 1964, Box 2324, NPS Records; Walter H. Kittams to John J. Craighead, 3 Mar. 1964, Box N-378, NPS Yellowstone Records.

26. John J. Craighead, "Wilderness in Montana . . . A Report," *Living Wilderness* 29 (Winter–Spring 1964): 24–29, on 28; "Problems of 'Wilderness Management' . . . ," *Living Wilderness* 29 (Autumn 1965): 37. On the Wilderness Act, see Roderick Nash, *Wilderness and the American Mind* (New Haven, CT: Yale University Press, 1967); Max Oelschlaeger, *The Idea of Wilderness: From Prehistory to the Age of Ecology* (New Haven, CT: Yale University Press, 1991).

27. According to Sellars, the creation of Sprugel's new division created tension

between wildlife rangers, with their more "traditional perspective on natural management," and scientists, who were "much more attuned to current ecological thinking"; Sellars, *Preserving Nature*, on 223; see also Pritchard, *Preserving Yellowstone's Natural Conditions*, 206.

28. John J. Craighead to Lemuel A. Garrison, 30 Nov. 1964, Garrison to Craighead, 29 Dec. 1964, Craighead to George B. Hartzog, 1964, Craighead to Walter H. Kittams, 30 Nov. 1964, Box N-378, NPS Yellowstone Records; Pritchard, *Preserving Yellowstone's Natural Conditions*, 239–241; Park Management Biologist to Chief Park Ranger, 5 May 1964, Craighead to John S. McLaughlin, 28 June 1965, Box N-91, NPS Yellowstone Records.

29. Adolph Murie to Freeman Tilden, 1 Jan. 1965, Box 2, Murie (Adolph) Files; Freeman Tilden, *The National Parks: What They Mean to You and Me* (New York: Knopf, 1951).

30. Adolph Murie to George Sprugel Jr., 11 Jan. 1965, Sprugel to Murie, 15 Jan. 1965, Box 2, Murie (Adolph) Files.

31. Richard Prasil to Adolph Murie, 18 Jan. 1965, Murie to Prasil, 20 Jan. 1965, Box 2, Murie (Adolph) Files. In April 1965, Murie stressed to the editor in chief of Houghton Mifflin that his grizzly monograph would include comments to the effect that "scientific studies that are harmful to the spirit of a park should not be permitted"; Murie to Paul Brooks, 25 Apr. 1965, Box 2, Murie (Adolph) Files.

32. Dean Interview; *Master Plan Brief for Mount McKinley National Park* (Anchorage, AK: National Park Service, 1965), on 5, 7; Richard G. Prasil, "Grizzly Bear Observations: Mount McKinley National Park, McKinley Park, Alaska, 1968," unpublished report, 2 pp., on p. 2, in the Alaska Resources Library and Information Services, Anchorage, AK. Alice Wondrak Biel, *Do (Not) Feed the Bears: The Fitful History of Wildlife and Tourists in Yellowstone* (Lawrence: University of Kansas Press, 2006), on 100.

33. J. M. Carpenter to Merrill P. Spencer, 26 July 1966, George Sprugel Jr., to Fred A. Glover, 22 July 1966, Box N-175, NPS Yellowstone Records. On Sprugel's resignation, see Sellars, *Preserving Nature*, 225–226. Oscar Godbout, "Animals and Fish Find It Hard to Keep Ahead of Scientific Advances," *New York Times*, 10 Feb. 1967.

34. Peter Arnold to Allan Morgan, 23 Feb. 1967, John J. Craighead to Walter E. Berlet, 8 Mar. 1967, Craighead to Carl Buchheister, 8 Mar. 1967, Craighead to Charles H. Callison, 20 Apr. 1967, John S. McLaughlin to Craighead, 14 Sept. 1967 and 29 Sept. 1967, Box N-176, NPS Yellowstone Records.

35. Walter E. Berlet to John J. Craighead, 25 May 1967, Craighead to Berlet, 31 May 1967, Box N-176, NPS Yellowstone Records.

36. John J. Craighead to Robert Pantzer, 20 Apr. 1967, Craighead to John S. McLaughlin, 28 June 1967, Box N-176, NPS Yellowstone Records. The *Los Angeles Times* notified its readers of the date of the broadcast in June 1967, more than three months in advance; "'Grizzly' Special Set Nov. 1," *Los Angeles Times*, 19 June 1967.

37. Yellowstone Monthly Report for June, 17 Aug. 1967, Box 104, NPS Records; John J. Craighead to John McLaughlin, n.d. [Summer 1967], Box N-176, NPS Yellowstone Records; Sellars, *Preserving Nature*, 250–252.

38. Craighead, *Track of the Grizzly*, on 195; Glen F. Cole, *Montana Deer Management: Where Do We Go from Here?* Montana Fish and Game Department Information Bulletin no. 1 (Helena, MT: Fish and Game Department, 1958), on 3; John S. McLaughlin to

John J. Craighead, 25 July 1967, Box N-92, NPS Yellowstone Records. See also Pritchard, *Preserving Yellowstone's Natural Conditions*, 224; Sellars, *Preserving Nature*, 231.

39. Superintendent, Yellowstone National Park [Jack Anderson], to Director, National Park Service [George Hartzog], 16 Aug. 1967, Box N-176, NPS Yellowstone Records; Research Biologist [Glen F. Cole] to Superintendent, Yellowstone [Jack Anderson], Re: "Comments on Craighead Report on Bear Management in Yellowstone National Park," 21 Aug. 1967, Box 2, Murie (Adolph) Files; Pritchard, *Preserving Yellowstone's Natural Conditions*, 243–245; Biel, *Do (Not) Feed the Bears*, 98; Glen F. Cole to Adolph Murie, 5 Dec. 1969, Box 2, Murie (Adolph) Files.

40. Adolph Murie to Michael Gawel, 24 Feb. 1969, Box 2, Murie (Adolph) Files.

41. Neil Bibler to John J. Craighead, 9 Sept. 1967, Craighead to Bibler, 22 Aug. 1967, Box N-176, NPS Yellowstone Records; *Grizzly!* videocassette, directed by Irwin Rosten (Washington, DC: National Geographic Society and Wolper Productions, 1967); Lawrence Laurent, "A Fine Study of 2 Naturalists," *Washington Post*, 2 Nov. 1967; Lois Crisler to Adolph Murie, n.d. [1967–1968], Box 1, Murie (Adolph) Files. On Crisler, see Gregg Mitman, *Reel Nature: America's Romance with Wildlife on Film* (Cambridge, MA: Harvard University Press, 1999), 116–118.

42. Supervisory Research Biologist [Glen F. Cole] to Superintendent [Jack Anderson], Yellowstone National Park, 16 Oct. 1967, Jack K. Anderson to Regional Director, Midwest Region, 6 Mar. 1968, Box N-172, NPS Yellowstone Records; Biel, *Do (Not) Feed the Bears*, 97–102; Peter T. Bromley to Anderson, 2 Dec. 1969, Anderson to Bromley, 19 Dec. 1969, Box N-196, NPS Yellowstone Records.

43. A. Starker Leopold to George Hartzog, 6 Oct. 1969, and attached report, "A Bear Management Policy and Program for Yellowstone National Park," Carton 5, Leopold Papers.

44. John J. Craighead to Jack K. Anderson, 14 Jan. 1969, Glen F. Cole to A. Starker Leopold, 29 Jan. 1969, Box N-92, NPS Yellowstone Records; "Robert M. Linn, 1926–2004: A Remembrance of the GWS Co-Founder," *George Wright Forum* 21 (2004): 4–9; Sellars, *Preserving Nature*, 227–228; Leopold to George B. Hartzog and John Gottschalk, 29 July 1968, Hartzog to Leopold, 5 Aug. 1968, Leopold to Sigurd F. Olson, Charles E. Olmsted, and Stanley Cain, 20 Aug. 1968, Box N-172, NPS Yellowstone Records; Supervisory Research Biologist [Glen Cole] to Superintendent [Jack Anderson], 12 Aug. 1968, Box N-118, NPS Yellowstone Records. Anderson later reiterated Cole's concern that highly conspicuous marking "could lead to restrictions on the use of markers as a research tool"; Anderson to John J. Craighead, 17 July 1969, Box N-91, NPS Yellowstone Records.

45. Jack K. Anderson to John J. Craighead, 7 Apr. 1969, Box N-91, NPS Yellowstone Records; Director, Midwest Region [Fred C. Fagergen] to Superintendent, Yellowstone National Park [Jack Anderson], 6 Mar. 1969, Box N-92, NPS Yellowstone Records.

46. John J. Craighead to Jack K. Anderson, 14 Apr. 1969, Box N-92, NPS Yellowstone Records.

47. A. Starker Leopold to George Hartzog, 6 Oct. 1969, and attached report, "A Bear Management Policy and Program for Yellowstone National Park," Glen F. Cole to Leopold, 10 Nov. 1969, Carton 5, Leopold Papers.

48. Craighead, *Track of the Grizzly*, on 165, 201–202. Craighead later claimed that Anderson told him the decision to remove tags had been recommended by the Park Service's Natural Science Advisory Committee, but that when Craighead asked Stanley Cain

in person if that was true at a meeting in September 1969 of the Natural Science Advisory Committee and park leadership, in which Fred Dean and Albert Erickson participated as outside consultants, he denied it. Anderson was forced, according to Craighead, to admit that it had been his own decision; Craighead, *Track of the Grizzly*, 202–204.

49. James Bradley, Assistant Secretary of the Smithsonian, Announcement Regarding Appointment of Sidney R. Galler, 9 Aug. 1965, Frank C. Craighead, "The Prospects for Using Nimbus B Satellite for Studying Large Land Animals," in *Some Prospects for Using Communications Satellites in Wild Animal Research* (Washington, DC: American Institute of Biological Sciences, 1966), on 4–5, Box 20, Smithsonian Record Unit 271.

50. Helmut K. Buechner, Grant Proposal to NASA, "Satellite (IRLS) Tracking of Elk in Yellowstone National Park and Caribou in Alaska," 21 Jan. 1969, Box 20, Smithsonian Record Unit 271; Russell E. Train to S. Dillon Ripley, 25 July 1969, Box N-196, NPS Yellowstone Records; Robert M. Linn to Glen F. Cole, 29 Oct. 1969, Box N-92, NPS Yellowstone Records; Buechner to George J. Jacobs, 16 Dec. 1969, Box 58, Buechner Papers; Ronald E. Lambertson to Jack K. Anderson, 24 Nov. 1969, Box N-196, NPS Yellowstone Records.

51. John J. Craighead et al., "Satellite Monitoring of Black Bear," *BioScience* 26 (1971): 1206–1212; Helmut K. Buechner, "First Progress Report on 'Satellite (IRLS) Tracking of Elk,'" 14 Oct. 1969, Helmut K. Buechner to George Jacobs, 12 Mar. 1970, John J. Craighead, Frank C. Craighead, Charles E. Cote, and Helmut K. Buechner, "Second Progress Report on 'Satellite (IRLS) Tracking of Elk,'" 8 Apr. 1970, Box 59, Buechner Papers.

52. Mary Hazel Harris, "'Monique' Death Ends Project," *Defenders of Wildlife News* (Mar. 1970): 19; Richard B. Selander, "Protest against Treatment of Elk," *Animal Welfare Institute Information Report* (Apr.–June 1970): 3; S. Dillon Ripley to Representative Charles E. Bennett, 11 Dec. 1970, Fred G. Evenden to Helmut K. Buechner, 5 Mar. 1970, Box 59, Buechner Papers. For examples of letters protesting the collaring of Monique, see Mrs. R. Thorn (Leola, PA) to Helmut K. Buechner, with attached letter from members of sixth-grade class, n.d. [Spring 1970], Marie Vance (Albany, NY) to S. Dillon Ripley, 29 June 1970, Dorothy Keyser (Waukegan, MI) to Buechner, 29 Oct. 1970, Mrs. S. N. Levens to Ripley, 26 Oct. 1970, Joanne J. Rongo (Providence, RI) to Buechner, 9 Nov. 1970, Box 58, Buechner Papers; Catherine E. Heppe to Ripley, 24 Nov. 1974, Box 20, Buechner Papers.

53. "Transcript of Proceedings: Presentation on Animal Tracking to Dr. Low, Washington," 4 June 1970, Box 59, Buechner Papers; Buechner to Walter Leuthold, 26 June 1972, Box 58, Buechner Papers; Charles E. Cote, interview by the author, Greenbelt, MD, 24 October 2007.

54. A. Starker Leopold to Jack K. Anderson, 22 June 1970, Anderson to Leopold, 26 June 1970, Box N-118, NPS Yellowstone Records; Anderson to John J. Craighead, 8 July 1970, Box N-91, NPS Yellowstone Records; Craighead to Anderson, 24 July 1970, Box N-92, NPS Yellowstone Records.

55. Edmund J. Bucknall to All Park Rangers through Chief Park Ranger, 6 Mar. 1971, Robert M. Linn to Raymond F. Dasmann, 21 Jan. 1971, Box N-116, NPS Yellowstone Records.

56. Edmund J. Bucknall to All Park Rangers thru Chief Park Ranger, 6 Mar. 1971, Box N-116, NPS Yellowstone Records.

57. John J. Craighead to Jack K. Anderson, 13 July 1971, Anderson to Craighead, 20 Aug. 1971, Box N-112, NPS Yellowstone Records.

58. Director, National Park Service [George Hartzog], to Director, Bureau of Sport Fisheries and Wildlife [John Gottschalk], 11 Aug. 1971, Robert M. Linn to Mike Mansfield, 3 Nov. 1972, Box N-36, NPS Yellowstone Records.

59. Robert B. Finley Jr. to Jack K. Anderson, 15 June 1973, Anderson to Finley, 27 June 1973, Box N-36, NPS Yellowstone Records; Anderson to Don Frickie, 22 Oct. 1974, Frickie to Anderson, 26 Sept. 1974, Box N-39, NPS Yellowstone Records. Historians Sellars and Pritchard dismiss concerns about tagging as trivial. Sellars describes the Yellowstone administration's attacks on wildlife tagging as an "excuse" for obstructing the Craigheads' research, thinly justified in terms of the Park Service's mandate to promote public enjoyment of the parks. Pritchard suggests that the conflict over tagging was one of several "minor irritations" magnified by personality conflict and explicitly endorses the use of radio collars, which "put the modern ecologist miles ahead of the natural historian's estimates of wildlife numbers and deductions about animal movements"; Sellars, *Preserving Nature*, 252; Pritchard, *Preserving Yellowstone's Natural Conditions*, 242. Biel, in contrast, points out that Cole recommended against using tags in Isle Royale National Park a few years later and that Anderson removed road signs and took other actions to make the appearance of Yellowstone more natural that had nothing to do with grizzlies or the Craigheads; Biel, *Do (Not) Feed the Bears*, 98.

60. Thomas McNamee, *The Grizzly Bear* (Guilford, CT: Globe Pequot, 1997), 117; Nathaniel P. Reed to Director, National Park Service, 10 Nov. 1973, Reed to Lee Metcalf, 19 Nov. 1974, Box N-119, NPS Yellowstone Records.

61. Craighead, *Track of the Grizzly*, 229; Biel, *Do (Not) Feed the Bears*, 113; Richard R. Knight, "Holding On to Yellowstone's Grizzlies: A Parting Chat with a 24-Year Veteran," *Yellowstone Science* 6 (1998): 2–9, on 3, also cited in Biel, *Do (Not) Feed the Bears*, 111; Richard Knight to Glen F. Cole, 23 Dec. 1975, Robert C. Haraden to Knight, 31 Dec. 1975, Box N-39, NPS Yellowstone Records; Pritchard, *Preserving Yellowstone's Natural Conditions*, 252.

62. On McKinley's wildlife management plan, see Assistant Director [Edward Hummel] to Regional Director, Northwest Region, 14 Jan. 1970, Box 7, NPS AK Task Force Files. In 1968, Dean recommended limiting research not directly related to conservation in Katmai National Monument; Frederick C. Dean, "Brown Bear/Human Interrelationship Study: Katmai," National Park Service Natural Resources Final Report AR-68-04 (Anchorage, AK: National Park Service, 1968). Dean later recalled that he had always been sympathetic to Murie's concerns, even if his judgment about the value of radio-tracking McKinley's bears had differed; Dean Interview.

63. Tom Adams [Chairman, Denali Citizens Council] to C. B. Harry, 28 Mar. 1976, C. B. Harry, "Assessment of the Environmental Impact on Proposed Study of the Distribution and Movement of the McKinley Caribou," 8 June 1976, Frederick C. Dean, "McKinley Caribou Herd, Research Planning Meeting," Notes and Summary, 6 Feb. 1976, Box 52, Alaska Region, Cultural Resource Study, Administrative Subject Files, 1966–1978, National Park Service, Record Group 79, National Archives and Records Administration, Anchorage, AK; Troyer, *Into Brown Bear Country*, xiii. Troyer later compared the shift from the wildlife research techniques of the late 1950s to those of the mid-1970s as "not unlike going from horse and buggy to modern cars." The new techniques were more effective and less stressful for men and bears alike, he wrote, making it possible

to conduct research "discreetly, with as little harassment as possible"; Troyer, *Into Brown Bear Country*, on 96–97.

64. Jan O. Murie, preface to *The Grizzlies of Mount McKinley*, by Adolph Murie (Washington, DC: Government Printing Office, 1981), xi–xii, on xii; Murie, *Grizzlies of Mount McKinley*, on 241.

65. Robert L. Rausch, review of *Grizzlies of Mount McKinley*, by Adolph Murie, *Journal of Wildlife Management* 47 (1982): 570–571, on 570; Charles Jonkel, review of *Grizzlies of Mount McKinley*, by Adolph Murie, *Journal of Mammalogy* 64 (1983): 187–188, on 188.

66. Craighead, *Track of the Grizzly*, 191–230; Maurice G. Hornocker, preface to *Definitive System for Analysis of Grizzly Bear Habitat and Other Wilderness Resources*, by J. J. Craighead, J. S. Sumner, and G. B. Scaggs (Missoula, MT: Wildlife-Wetlands Institute, 1982), vii. On early efforts to use satellite images to characterize grizzly habitat, see John Craighead, "Studying Grizzly Habitat by Satellite," *National Geographic* 150 (July 1976): 148–158.

67. Craighead, Sumner, and Scaggs, *Definitive System*, 206, also cited in Mitman, "When Nature *Is* the Zoo," 141; John J. Craighead and John A. Mitchell, "Grizzly Bear: Ursus Arctos," in Joseph A. Chapman and George A. Feldhamer, eds., *Wild Mammals of North America: Biology, Management, and Economics* (Baltimore: Johns Hopkins University Press, 1982), 515–556, on 549, 555.

68. Hank Fischer with Les Pengelly and John Craighead, "Voice for the Grizzlies," *Defenders* 58 (Nov.–Dec. 1983): 5–17, on 7, 9, 10–11; James Coates, "Drug for Capturing Grizzlies May Be Making Them Wilder," *Chicago Tribune*, 23 Sept. 1984.

69. L. David Mech, *The Wolf: The Ecology and Behavior of an Endangered Species* (Garden City, NY: Published for the American Museum of Natural History by Natural History Press, 1970); L. David Mech, *The Wolves of Denali* (Minneapolis: University of Minnesota Press, 1998), 5–6, 38.

70. Mech, *Wolves of Denali*, references to Murie on 3, 175, quotes on 27, 175. At least one reviewer of *The Wolves of Denali* followed Mech in describing the book as a "modern update of Murie's earlier work"; Victor Van Ballenberghe, review of *The Wolves of Denali*, by L. David Mech, *Journal of Wildlife Management* 63 (1999): 412–414, on 413. On the radio collaring of Denali's grizzlies, see Patricia Owen, "Grizzly Bear Population Ecology and Monitoring in Denali National Park and Preserve," *Arctic Research of the United States* 16 (2002): 18–21; Patricia A. Owen and Richard D. Mace, "Grizzly Bear Population Ecology in Denali National Park and Preserve," in *Alaska Park Science: Crossing Boundaries in a Changing Environment*, Proceedings of the Central Alaska Park Science Symposium, Sept. 12–14, 2006, www.nps.gov/akso/AKParkScience/symposium2006.

71. Mark Johnson, "Bugged Bears and Collared Cougars," *Yellowstone Science* 1 (Fall 1992): 6–9; L. David Mech and Shannon Barber, "A Critique of Wildlife Radio-Tracking and Its Use in National Parks: A Report to the U.S. National Park Service," U.S. National Park Service and U.S. Geological Survey, Northern Prairie Wildlife Research Center, Jamestown, ND, 2002, 78 pp., on p. 2. Asked about his experience with criticisms of radio collaring, Mech later recalled that "there's always some people who worry about . . . things like that, but all we worried about is, . . . is this going to be hard on the animal or is it going to interfere with the data we're collecting?" Mech Interview. For the results of Biel's survey, see Biel, *Do (Not) Feed the Bears*, 138–140.

72. In a 1985 article Alaskan historian Morgan Sherwood lamented, "Not even the

brown bear can count on roaming freely in his territory without being shot with a tranquilizer, tagged and equipped with a radio transmitter," but he neglected to mention the almost identical concerns that Murie had presented privately twenty years earlier, which had recently been published in his posthumous grizzly monograph. When *Saving America's Wildlife*, Thomas Dunlap's history of predator control in the United States, appeared in 1988, it credited Murie with helping to transform the Park Service's approach to predators but made no mention of his protests against tagging. In 2001, Timothy Rawson briefly mentioned Murie's skepticism toward radio tagging in *Changing Tracks*, his history of predator control in McKinley, only to largely dismiss those concerns as signs of the growing rigidity of a man nearing the end of his career. Morgan Sherwood, "The End of American Wilderness," *Environmental Review* 9 (1985): 197–209, on 203; Thomas R. Dunlap, *Saving America's Wildlife: Ecology and the American Mind, 1850–1990* (Princeton, NJ: Princeton University Press, 1988), on 176; Timothy Rawson, *Changing Tracks: Predators and Politics in Mt. McKinley National Park* (Fairbanks: University of Alaska Press, 2001), 263. On the dedication of the Murie Center, see Denali National Park and Preserve Superintendent's Annual Report, Fiscal Year 2004, www.nps.gov/dena/parkmgmt; Ingrid Nixon, "Science and Learning in the Alaska Wilderness," *International Journal of Wilderness* 11 (2005): 35–36, on 35. Another article by Denali staff suggested that Murie's studies of caribou from the late 1930s to 1965 were part of a seamless tradition of scientific research in the park that was continued by aerial tracking in the late 1960s and subsequently supplemented by "radio collaring, which is the technique used today." Despite quoting a passage from *The Grizzlies of Mount McKinley* in which Murie argued that "wild grizzlies in McKinley National Park, conducting their affairs undisturbed, are the essence of wilderness spirit," the article, like the exhibits, failed to mention Murie's vehement opposition to tagging and collaring or the serious discussions around Troyer's caribou-tracking proposal that were held in the mid-1970s; Ann Kain, "Scientific Legacy of Denali," *Alaska Park Science* 5 (2006): 59–61, on 60. The quoted Murie passage can be found in *Grizzlies of Mount McKinley*, xvi.

CHAPTER THREE: Diplomatic and Political Subtleties

1. S. Dillon Ripley to Daniel Patrick Moynihan, 18 Apr. 1973, and attached "Aide Memoire," Box 24, Smithsonian Record Unit 254, also cited in Michael Lewis, *Inventing Global Ecology: Tracking the Biodiversity Ideal in India, 1945–1997* (Hyderabad, India: Orient Longman, 2003), 271; Christen Wemmer, Ross Simons, and Hemanta Mishra, "Case History of a Cooperative International Conservation Program: The Smithsonian Nepal Tiger Ecology Project," 1984, unpublished manuscript, 32 pp., on p. 5, Box 59, Smithsonian Record Unit 329; Michael Lewis, "Indian Science for Indian Tigers? Conservation Biology and the Question of Cultural Values," *Journal of the History of Biology* 38 (2005): 185–207, on 193–194.

2. Roy MacLeod, "'Strictly for the Birds': Science, the Military and the Smithsonian's Pacific Ocean Biological Survey Program, 1963–1970," *Journal of the History of Biology* 34 (2001): 315–352; Lewis, *Inventing Global Ecology*, 132–142; Lewis, "Indian Science," 190; S. Dillon Ripley to J. W. Fulbright, 6 June 1969, Fulbright to Ripley, 17 June 1969, Sidney R. Galler to Ripley, 18 Dec. 1968, Box 10, Smithsonian Institution, Assistant Secretary for Science Records, 1963–1973, Record Unit 108, Smithsonian Institution Archives, Washington, DC.

3. Although biologists would later contest some of Hornocker's claims, at least one looked back on his work and that of his graduate student John C. Seidensticker as a key turning point when "the shroud around the 'mysterious American cat,' as it had once been called, was pulled away." Ken Alvarez, *Twilight of the Panther: Biology, Bureaucracy and Failure in an Endangered Species Program* (Sarasota, FL: Myakka River, 1993), on 66. Seidensticker Interview; Maurice G. Hornocker to Kennedy D. Schmertz, 23 July 1971, and attached "Report to Smithsonian Institution," Director, Ecology Program, to Director, Office of Environmental Sciences, 15 Sept. 1971, I. E. Wallen to Dale W. Jenkins, 16 Sept. 1971, Box 15, Smithsonian Record Unit 271.

4. On the Smithsonian's use of PL 480 funds and plans for research stations in India, see Lewis, *Inventing Global Ecology*, 93–95.

5. John C. Seidensticker IV and Maurice G. Hornocker, "Preliminary Proposal for Research on Ecology of the Tiger (Panthera tigris)," 5 Feb. 1972, Seidensticker to Michael Huxley, 17 Apr. 1972, and attached John C. Seidensticker, "Tiger Research Development Trip (Smithsonian Foreign Currency Program)," Mar. 1972, Box 15, Smithsonian Record Unit 271; Lewis, *Inventing Global Ecology*, 166.

6. Wemmer, Simons, and Mishra, "Case History," 6; Russell Train, *Politics, Pollution, and Pandas: An Environmental Memoir* (Washington, DC: Island, 2003), 280–282; Charles A. Lindbergh to Harold J. Coolidge, 29 Mar. 1969, Box 6, Series 78.16, Harold Jefferson Coolidge Papers, Harvard University Archives, Cambridge, MA.

7. Frank G. Nicholls to Zafar Futehally, 19 May 1972, and attached manuscript, Gerardo Budowski, "Scientific Imperialism," 11 Apr. 1972, on 1, 18, Box 110, Smithsonian Record Unit 254. Futehally would later urge Indian government officials not to throw away the "opportunity to harness the best talent of the world in the sphere of ecological studies," among whom he included Seidensticker, but his voice and those of other scientists at the Bombay Natural History Society would be ignored; see Lewis, *Inventing Global Ecology*, 167. S. Dillon Ripley to David Challinor, Michael Huxley, and Kennedy Schmertz, 30 May 1972, Box 110, Smithsonian Record Unit 254.

8. Wemmer, Simons, and Mishra, "Case History," 8. Sankhala's "Smithsonian laboratory" comment is quoted in Lewis, *Inventing Global Ecology*, 166. Seidensticker Interview; Kirti Man Tamang, Personal Resume, Box 53, Smithsonian Record Unit 329; John C. Seidensticker IV to Michael R. Huxley, 30 Aug. 1972, Box 24, Smithsonian Record Unit 254.

9. S. Dillon Ripley to Kennedy D. Schmertz, 12 Oct. 1972, Box 15, Smithsonian Record Unit 271. Lewis writes that "the Indian state's power to control its territory trumped American economic power to bribe its way into India's national parks, at least for a few years"; Lewis, *Inventing Global Ecology*, 237.

10. Seidensticker Interview; Mishra Interview; Kirti Man Tamang to Michael R. Huxley, 18 Oct. 1972, Tamang to Huxley, 24 Oct. 1972, Maurice G. Hornocker to Huxley, 8 Nov. 1972, Jitendra R. Sharma to S. Dillon Ripley, 4 Jan. 1973, Box 24, Smithsonian Record Unit 254.

11. Jitendra R. Sharma to S. Dillon Ripley, 4 Jan. 1973, Box 24, Smithsonian Record Unit 254; Seidensticker Interview; Hemanta R. Mishra with Jim Ottaway Jr., *Soul of a Rhino: A Nepali Adventure with Kings and Elephant Drivers, Billionaires, and Bureaucrats, Shamans and Scientists, and the Indian Rhinoceros* (Guilford, CT: Lyons, 2008), 70; see also Eric Dinerstein, *Tigerland and Other Unintended Destinations* (Washington, DC: Island, 2005), on 32.

12. Mishra Interview; John Blower to Michael R. Huxley, 11 Jan. 1973, Box 24, Smithsonian Record Unit 254.

13. On Ripley's continued hopes for India, see S. Dillon Ripley to Daniel Patrick Moynihan, 18 Apr. 1973, and attached "Aide Memoire," Box 24, Smithsonian Record Unit 254; Lewis, *Inventing Global Ecology*, 237. On WWF, see Ripley to Michael R. Huxley, 13 Sept. 1973, Box 120, Smithsonian Record Unit 254; S. Dillon Ripley, Memorandum for the Record, Re: "World Wildlife Fund: Reorganization of Administration of the American Appeal," 17 Mar. 1973, C. R. Gutermuth to All WWF Directors, 10 Apr. 1973, WWF-US, "Projects for Recommendation to the April 16, 1973 Board Meeting for Approval and Fund Raising," 16 Apr. 1973, Box 1, Series 78.19, Harold Jefferson Coolidge Papers, Harvard University Archives, Cambridge, MA.

14. John C. Seidensticker IV to Michael R. Huxley, 11 Apr. 1973, Box 24, Smithsonian Record Unit 254.

15. John C. Seidensticker to Leonard Carmichael, 20 Aug. 1973, Box 24, Smithsonian Record Unit 254; Seidensticker Interview.

16. Mishra, *Soul of a Rhino*, 82–86.

17. Mishra, *Soul of a Rhino*, 82–86. Charles McDougal to Secretary of Forests, His Majesty's Government of Nepal, 8 Sept. 1973, John H. Blower to Hemanta R. Mishra, 11 Sept. 1973, and attached handwritten note signed "FP" [Frank Poppleton], n.d., Box 24, Smithsonian Record Unit 254. Trained as an anthropologist, McDougal later developed a method of identifying individual tigers from photographs of their distinctive facial stripes; Charles McDougal, *The Face of the Tiger* (London: Rivington Books, 1977).

18. Seidensticker Interview; Emerald J. B. Rana to Michael R. Huxley, 19 Sept. 1973, WWF International to WWF-US, 25 Nov. 1973, John C. Seidensticker IV to Michael R. Huxley, 2 Dec. 1973, Box 24, Smithsonian Record Unit 254; Mishra Interview.

19. Mishra Interview; Mishra, *Soul of a Rhino*, 150; John C. Seidensticker IV to Michael R. Huxley, 2 Dec. 1973, Box 24, Smithsonian Record Unit 254.

20. Mishra Interview; Seidensticker Interview; George B. Schaller, *The Deer and the Tiger: A Study of Wildlife in India* (Chicago: University of Chicago Press, 1967).

21. Seidensticker Interview; John Seidensticker, Kirti Man Tamang, and Clinton W. Gray, "The Use of CI-744 to Immobilize Free-Ranging Tigers and Leopards," *Journal of Zoo Animal Medicine* 5 (1974): 22–25.

22. Jim Doherty and Maurice Hornocker, "Cougars Up Close," *National Wildlife* 14 (1976): 43–47; Mishra Interview; Melvin E. Sunquist, *The Social Organization of Tigers (Panthera tigris) in Royal Chitawan National Park, Nepal* (Washington, DC: Smithsonian Institution Press, 1981), on 3.

23. John C. Seidensticker to Michael R. Huxley, 2 Dec. 1973, Seidensticker to Huxley, 25 Dec. 1973, Box 24, Smithsonian Record Unit 254; John C. Seidensticker and Kirti M. Tamang, "Background Information for the Time-Life Science Yearbook: Smithsonian Institution: Nepal Tiger Ecology Project," unpublished manuscript, 1 July 1974, on 1, Box 53, Smithsonian Record Unit 329.

24. S. Dillon Ripley to David Challinor, Godfrey Rockefeller, and Kirti M. Tamang, 8 Mar. 1976, Box 53, Smithsonian Record Unit 329; John C. Seidensticker IV to Michael R. Huxley, 19 Oct. 1973, Box 24, Smithsonian Record Unit 254.

25. Michael R. Huxley to John C. Seidensticker IV, 9 July 1974, Box 53, Smithsonian Record Unit 329.

26. Seidensticker Interview; Lewis, *Inventing Global Ecology*, 168–169.

27. Ross Simons to Kirti Man Tamang, 3 Sept. 1974, Box 53, Smithsonian Record Unit 329; Sunquist, *Social Organization*, 14; Fiona Sunquist and Melvin E. Sunquist, *Tiger Moon* (Chicago: University of Chicago Press, 1988), on 3–4.

28. David Challinor to S. Dillon Ripley, 16 Dec. 1974, Box 18, Smithsonian Record Unit 254; *World Wildlife Fund Annual Report 1975* (Morges, Switzerland: WWF, 1975), 11.

29. Seidensticker Interview. The quote on the "rigid system of territorialism" is in Doherty and Hornocker, "Cougars Up Close," on 46; see also Maurice G. Hornocker, "An Analysis of Mountain Lion Predation upon Mule Deer and Elk in the Idaho Primitive Area," *Wildlife Monographs* 21 (1970): 3–39. The term "land tenure system" is used in Schaller, *Deer and the Tiger*, on 241. Hornocker's work with the Craigheads on grizzlies in Yellowstone had similarly focused on dominance and territoriality; Craighead, *Track of the Grizzly*, 33, 142. J. Seidensticker, M. G. Hornocker, W. V. Wiles, and J. P. Messick, "Mountain Lion Social Organization in the Idaho Primitive Area," *Wildlife Monographs* 35 (1973): 1–60. The term "territorial imperative" was popularized in Robert Ardrey, *The Territorial Imperative: A Personal Inquiry into the Animal Origins of Property and Nations* (New York: Atheneum, 1966), which drew heavily on Helmut Beuchner's research on the territorial behavior of Ugandan kob. For Eisenberg's influence on Seidensticker, see John Seidensticker to Michal Huxley, 25 Feb. 1974, Box 53, Smithsonian Record Unit 329; D. G. Kleiman and J. F. Eisenberg, "Comparisons of Canid and Felid Social Systems from an Evolutionary Perspective," *Animal Behaviour* 21 (1973): 637–659, on 653.

30. Sunquist, *Social Organization*, 40–41, 43, 45, 49, 53, 61–62.

31. Mishra Interview; Peter A. Jordan to Michael R. Huxley, 29 Oct. 1975, Siniff Papers; J. L. David Smith to Ross Simons, n.d. [Feb. 1976], Box 53, Smithsonian Record Unit 329. The Dakre Tiger's unexpectedly large range eventually proved his undoing; in August 1976, he was found poisoned near the park border; Sunquist and Sunquist, *Tiger Moon*, 161; Sunquist, *Social Organization*, 39.

32. Wemmer, Simons, and Mishra, "Case Study," 18; J. L. David Smith to Ross Simons, n.d. [Feb. 1976], Box 53, Smithsonian Record Unit 329.

33. Hemanta R. Mishra to Charles McDougal, 5 Mar. 1976, J. L. David Smith to Ross Simons, n.d. [Feb. 1976], Box 53, Smithsonian Record Unit 329.

34. S. Dillon Ripley to David Challinor, Godfrey Rockefeller, and Kirti M. Tamang, 8 Mar. 1976, Box 53, Smithsonian Record Unit 329.

35. S. Dillon Ripley to David Challinor, Godfrey Rockefeller, and Kirti M. Tamang, 8 Mar. 1976, Box 53, Smithsonian Record Unit 329.

36. S. Dillon Ripley to David Challinor, 2 Apr. 1976, and attached article, Subhash Kirpekar, "Counting Tigers in the Wild," *Times of India*, 3 Mar. 1976, Box 53, Smithsonian Record Unit 329.

37. N. D. Jayal to S. Dillon Ripley, 26 Aug. 1976, and attached report, Project Tiger, "Report on the Training Course in Tranquilising and Telemetry, 6–23 April 1976," Box 27, Smithsonian Record Unit 254.

38. Kailash Sankhala, "Project Tiger," in K. Ullas Karanth, ed., *Tiger Tales: Tracking the Big Cat across Asia* (New Delhi: Penguin Books, 2006), 180–187, on 187; "Rescuing Wildlife," in *World Wildlife Fund Annual Report 1977* (Morges, Switzerland: WWF, 1977), unpaginated, 2 pp.

39. Donald B. Siniff to Ross Simons, 14 Apr. 1976, Siniff Papers; Chris Wemmer,

"Report on the Nepal Trip of 17 Jan.–14 Feb., 1977," n.d. [1977], Box 27, Smithsonian Record Unit 254.

40. Melvin E. Sunquist to Ross Simons, 29 Apr. 1976, Kirti Man Tamang to S. Dillon Ripley, 30 Apr. 1976, Box 53, Smithsonian Record Unit 329.

41. Melvin E. Sunquist to Ross Simons, 29 Apr. 1976, Kirti Man Tamang to S. Dillon Ripley, 30 Apr. 1976, Tamang to Ripley, 1 June 1976, Simons to Ripley through David Challinor, 10 June 1976, Box 53, Smithsonian Record Unit 329.

42. Ross Simons to Chris Wemmer, 31 Aug. 1976, Box 27, Smithsonian Record Unit 254.

43. "Rescuing Wildlife," in *World Wildlife Fund Annual Report 1977*; Ross Simons to Chris Wemmer, 31 Aug. 1976, Box 27, Smithsonian Record Unit 254.

44. Emerald J. B. Rana to S. Dillon Ripley, 27 Sept. 1976, Ross Simons to Kirti Man Tamang, J. L. David Smith, and Rebecca Troth, 10 Nov. 1976, Box 53, Smithsonian Record Unit 329. On WWF International's renewed opposition, see Duncan Poore to Charles de Haes, 3 Sept. 1976, Fritz Vollmar to Godfrey Rockefeller, 26 Aug. 1976, Box 53, Smithsonian Record Unit 329; Simons to Chris Wemmer, 31 Aug. 1976, Box 27, Smithsonian Record Unit 254. Peter A. Jordan to David Challinor, 14 Dec. 1976, Box 27, Smithsonian Record Unit 254. On Smith and Jordan's interest in Sudan, see El Rayah O. Hassaballa to Peter A. Jordon [Jordan], 1 Mar. 1977, Jordan to Hassaballa, 18 Mar. 1977, Siniff Papers.

45. David Challinor and Ross Simons to S. Dillon Ripley, 3 Apr. 1977, Birendra Bahadur Shah to Challinor, 4 Oct. 1977, Box 53, Smithsonian Record Unit 329; "Rescuing Wildlife," in *World Wildlife Fund Annual Report 1977*; J. L. David Smith to Christen Wemmer, 7 June 1977, Box 7, Smithsonian Acc. 03-037; see also Wemmer, Simons, and Mishra, "Case Study," 19–20.

46. Peter A. Jordan, "Trip Report: Visit to Smithsonian Tiger Project, Royal Chitwan National Park," Nov. 1977, Box 54, Smithsonian Record Unit 329; J. L. David Smith to Christen Wemmer, 7 June 1977, Box 7, Smithsonian Acc. 03-037.

47. Douglas Heck to S. Dillon Ripley, 31 Oct. 1977, F. L. Kellogg to the WWF Director General, "Report on a Visit to Nepal, October 27–November 2, 1977," n.d., Heck to Kellogg, 15 Mar. 1978, David Challinor to Heck, 30 Mar. 1978, Box 54, Smithsonian Record Unit 329.

48. Challinor's recommendation letter was addressed to Frederick Dean at the University of Alaska, whose efforts to radio-tag the grizzlies of Mount McKinley National Park in the early 1960s had foundered on opposition from Adolph Murie and the McKinley superintendents; David Challinor to F. C. Dean, 13 Mar. 1985, Box 58, Smithsonian Record Unit 329. J. L. David Smith to Ross Simons, 31 Dec. 1977, Box 54, Smithsonian Record Unit 329; Douglas Heck to S. Dillon Ripley, 30 May 1978, Hemanta R. Mishra to Ross Simons, 25 June 1978, Box 55, Smithsonian Record Unit 329. Mishra's recollections of Smith's personality are from Mishra Interview; see also Mishra, *Soul of a Rhino*, 97.

49. Hemanta R. Mishra to Ross Simons, n.d. [1978], Box 55, Smithsonian Record Unit 329; Simons to Nancy Seaman, n.d. [1978], Box 53, Smithsonian Record Unit 329; Peter F. R. Jackson, "Scientists Hunt the Bengal Tiger—But Only in Order to Trace and Save It," *Smithsonian* 9 (Aug. 1978): 28–37.

50. Ross Simons to Richard Reinauer, 17 Apr. 1978, Box 54, Smithsonian Record Unit 329; Reinauer to Simons, 27 Apr. 1978, Simons to Mishra, 16 Nov. 1979, Box 55,

Smithsonian Record Unit 329; Simons to S. Dillon Ripley, 16 Apr. 1980, John Wilcox to Simons, 24 Apr. 1980, Box 56, Smithsonian Record Unit 329.

51. Mishra Interview; Hemanta R. Mishra to Colin Platt, 24 Oct. 1980, Russell E. Train to Virginia R. Bacher (Salinas, California), 21 Apr. 1981, Box 56, Smithsonian Record Unit 329. In April 1981, Ripley, in response to claims that the tiger project had been funded primarily by WWF, noted that the Smithsonian had provided more than half the funds for the project and that the contributions of Survival Anglia and ABC had been "significant"; S. Dillon Ripley to Guy Mountfort, 30 Apr. 1981, Box 56, Smithsonian Record Unit 329.

52. Biswa N. Upreti to Peter Burleigh, 29 June 1981, Burleigh to David Challinor, 17 July 1981, John Wilcox to Upreti, 23 July 1981, Box 56, Smithsonian Record Unit 329; Elizabeth Overmyer to Ross Simons, 21 June 1982, Box 57, Smithsonian Record Unit 329.

53. James L. David Smith to Ross Simons, 27 Jan. 1978, Theodore H. Reed to Smith, 14 Feb. 1978, Reed to Iain Taylor, 14 Feb. 1978, Box 54, Smithsonian Record Unit 329. Mishra himself was concerned about the amount of time he was forced to spend hosting visitors to Chitwan; see Hemanta R. Mishra to Chief, National Parks and Wildlife Office [Biswa Upreti], 12 Feb. 1979, Box 55, Smithsonian Record Unit 329.

54. Hemanta R. Mishra, *Nature Conservation in Nepal: An Introduction to the National Parks and Wildlife Conservation Programme of His Majesty's Government* (Kathmandu: Tribhuvan University Press, 1974), iii; Mishra, *Soul of a Rhino*, 77; Hemanta R. Mishra, "Balancing Human Needs and Conservation in Nepal's Royal Chitwan Park," *Ambio* 11 (1982): 246–251, on 247. Tiger attacks are described in Pralad Yonzon, "Human Tiger Interaction in Chitawan," in *International Symposium on Tigers, India, February 22–24, 1979: Papers, Proceedings and Resolutions* (New Delhi: Project Tiger, 1984), 126–128, on 128; Hemanta R. Mishra, Chris Wemmer, and J. L. D. Smith, "Tigers in Nepal: Management Conflicts with Human Interests," in Ronald L. Tilson and Ulysses S. Seal, eds., *Tigers of the World: The Biology, Biopolitics, Management, and Conservation of an Endangered Species* (Park Ridge, NJ: Noyes Publications, 1987), 449–463, on 455.

55. James L. David Smith, "Report to the Office National Parks and Wildlife Conservation: A Recent Human-Tiger Conflict Resulting in the Death of a Man," 1 Dec. 1979, Box 55, Smithsonian Record Unit 329; Mishra, Wemmer, and Smith, "Tigers in Nepal," 458; Mishra, "Balancing Human Needs," on 248.

56. J. L. David Smith to Ross Simons, 27 Jan. 1978, Box 54, Smithsonian Record Unit 329; Hemanta R. Mishra to Simons, 3 May 1979, and attached "Gentlemen's Agreement," 6 Nov. 1978, Box 55, Smithsonian Record Unit 329.

57. James L. David Smith to Francine Berkowitz, 10 Dec. 1982, Box 57, Smithsonian Record Unit 329; S. Dillon Ripley to Guy Mountfort, 13 Sept. 1979, Box 55, Smithsonian Record Unit 329; Charles McDougal, "Some Observations on the Social Behaviour of Tigers at Royal Chitawan National Park in Nepal," in *International Symposium on Tigers*, 113–118, on 118.

58. S. Dillon Ripley to Russell E. Train, 8 Nov. 1977, Box 3, Train Papers. The quotes on Train's plans for WWF-US are from Train, *Politics*, 244, 242.

59. David Challinor to Russell E. Train, 25 July 1979, Box 55, Smithsonian Record Unit 329. Mishra later claimed that the Smithsonian project's findings had resulted in nearly doubling the size of the park, from 544 square kilometers in 1973 to 1,040 square kilometers, including a new western extension established in 1978; Mishra, "Balancing

Human Needs," 246. As Lewis has pointed out, radiotelemetry studies of large carnivores produced data well suited to justifying the establishment of large nature reserves; they were not neutral contributors to the so-called SLOSS debate among conservation biologists over whether a "single large" or "several small" reserves were better for conservation; Lewis, *Inventing Global Ecology*, 274. Ross Simons to S. Dillon Ripley through David Challinor, 13 Jan. 1978, Box 54, Smithsonian Record Unit 329; Chris Wemmer to Simons, 3 May 1979, Box 55, Smithsonian Record Unit 329.

60. Biswa N. Upreti to Ross Simons, 28 Nov. 1979, Box 55, Smithsonian Record Unit 329; Douglas Heck to the Files, 19 Jan. 1980, Re: "Smithsonian Institution's Plans for Nepal," Box 56, Smithsonian Record Unit 329.

61. David Challinor to Prince Gyanendra Bir Bikram Shah, 15 Feb. 1980, Box 56, Smithsonian Record Unit 329; Train, *Politics*, on 283.

62. James L. David Smith, "Nepal, Project 1051, Tiger—Study of Biology in Chitwan National Park, WWF Grant 1980—$21,168," in *WWF International Yearbook 1980–1981* (Morges, Switzerland: WWF, 1981), 341–344; Hemanta R. Mishra to Thomas Lovejoy, 15 Aug. 1980, Box 56, Smithsonian Record Unit 329; Chris Wemmer to Charles McDougal, 9 Feb. 1982, Prem Bahadur Rai to Wemmer, 8 Mar. 1982, Box 57, Smithsonian Record Unit 329.

63. Russell E. Train to Ambassador Laise, 11 Mar. 1987, Box 1, Train Papers. The account of Chitwan in Train's memoir combines episodes from several Nepal visits; Train, *Politics*, 280–287. His contemporary account of the events can be found in Russell E. Train, 1981 Journal, entries for 6–9 Feb. 1981, Box 32, Train Papers. The quoted passage is in the entry for Feb. 7.

64. Hemanta R. Mishra to Kathryn S. Fuller, 18 July 1994, Mishra to Russell E. Train, 1 Aug. 1994, Box 13, Train Papers. A contemporary description of the Sauraha camp's tiger-darting puja is given in Hemanta Mishra to Chris Wemmer, 14 July 1981, Box 56, Smithsonian Record Unit 329. In 1983, one of the Nepalese managers of Tiger Tops reported that the lodge's shikaris and elephant drivers conducted two pujas each year, one in the summer and one in the winter, and commented that "an outsider might think the occasion lighthearted, but in fact it has the serious purpose of protecting those who work at Tiger Tops against wild animals"; K. K. Gurung, *Heart of the Jungle: The Wildlife of Chitwan, Nepal* (London: André Deutsch in association with Tiger Tops, 1983), on 167. For the description of the silk cotton tree blooms, see Russell E. Train, 1981 Journal, entry for 8 Feb. 1981, Box 32, Train Papers.

65. Russell E. Train, 1981 Journal, entry for 7 Feb. 1981, Box 32, Train Papers.

66. Ross Simons to Record, Re: "Meeting with AID—Nepal Director Brennan," 28 Oct. 1983, Box 57, Smithsonian Record Unit 329; Russell E. Train to Ambassador Laise, 11 Mar. 1987, Box 1, Train Papers; Train, *Politics*, 286–287.

67. "Statement by His Royal Highness Prince Gyanendra Bir Bikram Shah," 23 May 1984, Box 8, Smithsonian Acc. 03-037; Lewis, *Inventing Global Ecology*, 144; Hemanta R. Mishra to Ross Simons, 21 July 1984, Box 7, Smithsonian Acc. 03-037.

68. Chris Wemmer to David Challinor, "Trip Report: Nepal and India February 17–March 19, 1984," n.d., Box 59, Smithsonian Record Unit 329; Mishra, *Soul of a Rhino*, on 200; Chris Babu [Christen Wemmer] to Eric Bahadur Dinerstein and Hemanta Erasmus Mishra, 10 Dec. 1984, Box 58, Smithsonian Record Unit 329; Dinerstein, *Tigerland*, 13–15.

69. Eric Dinerstein to Chris Wemmer, 3 Aug. 1985, Dinerstein to Wemmer, 22 Aug. 1985, Box 58, Smithsonian Record Unit 329.

70. Hemanta R. Mishra to Ross Simons, 8 Aug. 1985, Box 7, Smithsonian Acc. 03-037, quoted verbatim.

71. Eric [Dinerstein] to Chris [Wemmer], 18 Nov. 1985, Box 7, Smithsonian Acc. 03-037; Dinerstein, *Tigerland*, on 266; Mishra, *Soul of a Rhino*, on 200.

72. On the numbers of rhinoceroses darted and collared, see Eric Dinerstein, Sunder Shrestha, and Hemanta Mishra, "Capture, Chemical Immobilization, and Radio-Collar Life for Greater One-Horned Rhinoceros," *Wildlife Society Bulletin* 18 (1990): 36–41, on 40. For the quote on trained elephants and their drivers, see Eric Dinerstein, *The Return of the Unicorns: The Natural History and Conservation of the Greater One-Horned Rhinoceros* (New York: Columbia University Press, 2003), on 65. On Philip's 1961 and 1986 visits to Chitwan, see John MacKenzie, *The Empire of Nature: Hunting, Conservation, and British Imperialism* (New York: Manchester University Press, 1988), on 310. On Train's 1987 visit, see Train, *Politics*, 285; Russell E. Train, 1987 Journal, entry for 20 Feb. 1987, Box 32, Train Papers.

73. Ross Simons to Hemanta R. Mishra, 28 July 1989, Box 18, Office of International Relations, Smithsonian Institution, Accession 01-114, Smithsonian Institution Archives, Washington, DC; Prince Gyanendra Bir Bikram Shah to Russell E. Train, 28 May 1990, Train to Prince Gyanendra, 30 May 1990, Train to Sir Arthur Norman, 9 Aug. 1990, Box 1, Train Papers; Mishra, *Soul of a Rhino*, on 149.

74. Mech Interview; L. David Mech and U. S. Seal, "Animal Immobilisation Film," in *World Wildlife Fund Yearbook 1982* (Gland, Switzerland: WWF International, 1982), 346. Mech's experiences in India inspired him to write the first general introduction to wildlife radio tracking; L. David Mech, *Handbook of Animal Radio-Tracking* (Minneapolis: University of Minnesota Press, 1983). On the Wildlife Institute of India, see Lewis, *Inventing Global Ecology*, 248, 253, 271.

75. Notes on contributors, in Karanth, *Tiger Tales*, 289–296, on 293; K. Ullas Karanth, "A View from the Machan," in *A View from the Machan: How Science Can Save the Fragile Predator* (Delhi: Permanent Black, 2006), 1–9.

76. K. Ullas Karanth "Understanding Tigers," in *View from the Machan*, 52–59, on 54. An earlier version of the article did not mention the technique's origins in big game hunting; K. Ullas Karanth "Understanding Tigers," *Wildlife Conservation* 98 (May–June 1995): 26–37, 74. R. Sukumar, "The Nagarhole Tiger Controversy," *Current Science* 59, no. 23 (10 Dec. 1990): 1213–1216, on 1215. On Sukumar's contribution to the turn away from strict "preservationist" approaches to wildlife conservation in India, see Mahesh Rangarajan, "The Politics of Ecology: The Debate on Wildlife and People in India, 1970–95," *Economic and Political Weekly* 31 (Sept. 1996): 2391–2394, 2396–2397, 2399, 2401–2404, 2406–2409, on 2394.

77. Lewis, "Indian Science," 202; K. Ullas Karanth and James D. Nichols, "Estimation of Tiger Densities in India Using Photographic Captures and Recaptures," *Ecology* 79 (1998), 2852–2862, on 2853; Fiona Sunquist, "Caught in the Trap!" *International Wildlife* 27 (Nov.–Dec. 1997): 40–47, on 40; Christen Wemmer, "Camera Trap Pioneers: Charles McDougal," 13 Feb. 2009, cameratrapcodger.blogspot.com.

78. Mel Sunquist and Fiona Sunquist, "Field Research Techniques: Recent Advances," in *Wild Cats of the World* (Chicago: University of Chicago Press, 2002), 397–399, quotes on 397, 398, 399.

CHAPTER FOUR: The Regulatory Leviathan

1. W. E. Evans and W. W. Sutherland, "Potential for Telemetry in Studies of Aquatic Animal Communication," in Lloyd E. Slater, ed., *Bio-Telemetry: The Use of Telemetry in Animal Behavior and Physiology in Relation to Ecological Problems* (New York: Pergamon, 1963), 217–224, on 220–221, 224. On the U.S. Navy's interest in cetaceans, see "Brief History of the Navy's Marine Mammal Program," in *Annotated Bibliography of Publications from the U.S. Navy's Marine Mammal Program*, Technical Document 627 (San Diego, CA: Space and Naval Warfare Systems Center, 1998): v–vii; Gregg Mitman, *Reel Nature: America's Romance with Wildlife on Film* (Cambridge, MA: Harvard University Press, 1999), 172.

2. Paul M. Fye to Carl L. Hubbs, 1 July 1964, William E. Schevill, "Application of Satellites or High-Flying Aircraft to Studies of Cetaceans and Other Large Marine Animals," paper presented at the Conference to Study the Feasibility of Conducting Oceanographic Explorations from Aircraft, Manned Orbital and Lunar Laboratories, Woods Hole Oceanographic Institution, 18 Aug. 1964, Box 24, Hubbs Papers; William A. Watkins and William E. Schevill, *Development and Testing of a Radio Whale Tag*, Woods Hole Oceanographic Institution Technical Report 77–58 (Woods Hole, MA: Woods Hole Oceanographic Institution, 1977).

3. On Adams's gray whale proposal, see Lowell Adams to Director, Fish and Wildlife Service, 25 June 1965, Adams to Carl L. Hubbs, 27 July 1965, Box 56, Hubbs Papers. On Hubbs's warning to Adams, see Hubbs to Adams, 10 Sept. 1965, Box 56, Hubbs Papers. On Hubbs's concern about "excess disturbance," see Hubbs to John Kanwisher, 17 Jan. 1966, Box 56, Hubbs Papers.

4. Stephen Leatherwood and William E. Evans, "Some Recent Uses and Potentials of Radiotelemetry in Field Studies of Cetaceans," in Howard E. Winn and Bori L. Olla, eds., *Behavior of Marine Animals: Current Perspectives in Research*, vol. 3 (New York: Plenum, 1979),: 1–31, on 3; George A. Llano to Roger S. Payne, 6 May 1968, Llano to Payne, 6 Nov. 1969, Box 7 and Box 10, Office of Antarctic Programs, Records of the Program Director for Biology and Medicine (G. Llano), General Correspondence and Related Records, 1961–1969, Entry 33, National Science Foundation, Record Group 370, National Archives and Records Administration, College Park, MD; Llano to Jack Schultz, 24 Apr. 1970, Box 1, Office of Polar Programs, Records of the Polar Science Section, Records of the Program Manager, Reading File, Entry 100, National Science Foundation, Record Group 370, National Archives and Records Administration, College Park, MD.

5. "Biographical Sketch of William Eugene Evans," in *Hearing before the Committee on Commerce, Science, and Transportation, United States Senate, on Nomination of Dr. William Evans to Be Chairman, Marine Mammal Commission, 21 March 1984*, 98th Cong., 2nd sess. (Washington, DC: Government Printing Office, 1984), 13–14. Like Evans, Norris had attended the 1962 biotelemetry conference; Kenneth S. Norris, "Preparations for Radiotelemetry of the Body Temperatures of Large Reptiles," in Slater, *Bio-Telemetry*, 283–287. On Evans's radiotelemetry research, see William E. Evans, "Orientation Behavior of Delphinids: Radio Telemetric Studies," *Annals of the New York Academy of Sciences* 188 (1971): 142–160; Leatherwood and Evans, "Some Recent Uses and Potentials of Radiotelemetry," 3; William E. Evans, *Fifty Years of Flukes and Flippers: A Little*

History and Personal Adventures with Dolphins, Whales and Sea Lions (1958–2007) (Sofia: Pensoft, 2008), 71.

6. Clark A. Bowers and R. Scott Henderson, *Project Deep Ops: Deep Object Recovery With Killer and Pilot Whales* (San Diego, CA: Naval Undersea Center, 1972), on 2–4, 53; G. W. Ching and H. O. Porter, *Project Deep Ops Equipment Development* (San Diego, CA: Naval Undersea Center, June 1972), on 8; "Brief History of the Navy's Marine Mammal Program," in *Annotated Bibliography of Marine Mammal Program*, v–vii, on vii. On the 1966 proposal to NSF, which was submitted by engineer Howard Baldwin, see Howard A. Baldwin to Philip M. Smith, 19 May 1966, Box 5, Office of Antarctic Programs, Records of the Program Director for Biology and Medicine (G. Llano), General Correspondence and Related Records, 1961–1969, Entry 33, National Science Foundation, Record Group 370, National Archives and Records Administration, College Park, MD.

7. William E. Schevill, preface to William E. Schevill, G. Carleton Ray, and Kenneth S. Norris, eds., *The Whale Problem: A Status Report* (Cambridge, MA: Harvard University Press, 1974): v–viii, on v; W. E. Evans, "Radio-Telemetric Studies of Two Species of Small Odontocete Cetaceans," in Schevill et al., *Whale Problem*, 385–394; Kenneth S. Norris, W. E. Evans, and G. Carleton Ray, "New Tagging and Tracking Methods for the Study of Marine Mammal Biology and Migration," in Schevill et al., *Whale Problem*, 395–410, on 395.

8. George A. Bartholomew et al., "Report of the Working Group on Biology and Natural History," in Schevill et al., *Whale Problem*, 5–10; Conference Management Committee et al., "A Digest of the Major Conclusions of the International Conference on the Biology of Whales, 10–12 June 1971, Shenandoah National Park, Virginia," in Schevill et al., *Whale Problem*, 3–4, on 4. For a later argument that telemetry was essential to uncoupling whale science from whaling, see "Report of the Scientific Committee," 27 June 1973, IWC/25/4(Rev. 2), and attached annex, H: W. E. Schevill and K. Radway Allen, "Expanded Cetacean Research," 2 pp., on 1–2, International Whaling Commission Archives, National Marine Mammal Laboratory, National Marine Fisheries Service, Seattle, WA.

9. Carl L. Hubbs to Alan Cranston, 12 May 1971, Hubbs to Kenneth S. Norris, 13 May 1971, Box 17, Hubbs Papers. On the origins of the MMPA, see Mark V. Barrow, *Nature's Ghosts: Confronting Extinction from the Age of Jefferson to the Age of Ecology* (Chicago: University of Chicago Press, 2009), 335–336.

10. *Hearings on Legislation for the Preservation and Protection of Marine Mammals before the House Committee on Merchant Marine and Fisheries, Subcommittee on Fisheries and Wildlife Conservation*, 92nd Cong., 1st sess. (Washington, DC: U.S. Government Printing Office, 1971), on 79.

11. Roderick Frazier Nash, *The Rights of Nature: A History of Environmental Ethics* (Madison: University of Wisconsin Press, 1989), on 173; *Hearings on Legislation for the Preservation and Protection of Marine Mammals*, 1971, Schevill, Norris, and Ray's prepared statement on 399–406, quote on necessity of management on 402, Norris's statement on tracking on 428; International Conference on the Biology of Whales, Skyland, Shenandoah National Park, Luray, Va., June 10–12, 1971, Reports of the Working Groups, in *Hearings on Legislation for the Preservation and Protection of Marine Mammals*, 1971, 428–435.

12. For statements of support for Gigi's capture, see Carl L. Hubbs to David W. Kenney, 7 Jan. 1971, Robert Elsner to Kenney, 2 Nov. 1970, J. L. McHugh to Kenney, 26 Jan.

1971, Box 56, Hubbs Papers. On Wilkie's skepticism, see Donald W. Wilkie to Kenney, 26 Mar. 1971, Box 56, Hubbs Papers.

13. William E. Evans to Distribution, Re: "Release and Subsequent Radio Tracking of a Yearling California Gray Whale *Eschrichtius robustus*," 31 Jan. 1972, Charles H. Lawrence to Sea World, 12 Jan. 1972, Box 56, Hubbs Papers.

14. William E. Evans to Distribution, Re: "Release and Subsequent Radio Tracking of a Yearling California Gray Whale *Eschrichtius robustus*," 31 Jan. 1972, Box 56, Hubbs Papers.

15. On the results of tracking Gigi, see W. E. Evans, "Telemetering of Temperature and Depth Data from a Free Ranging Yearling California Gray Whale, *Eschrichtius Robustus*," *Marine Fisheries Review* 36 (Apr. 1974): 52–58; Carl L. Hubbs to Alan Baldridge, 9 June 1972, Box 53, Hubbs Papers. For Evans's later recollection of the public relations purposes of tagging, see Evans, *Fifty Years*, 87. The "baby whale" quote is from the subtitle of a youth nonfiction book later coauthored by Evans; Eleanor Coerr and William E. Evans, *Gigi: A Baby Whale Borrowed for Science and Returned to the Sea* (New York: Putnam, 1980).

16. "Death for Dolphins," *New York Times*, 9 Apr. 1972. The celebration of the MMPA as "a scientific management authority rather than an emotional reaction piece" by one wildlife manager was typical; Daniel Poole to Ira N. Gabrielson, 21 Aug. 1972, Box 1, Ira N. Gabrielson Papers, CONS 37, Denver Public Library, Denver, CO. Nixon's signing is mentioned in "Law Aids Natives in Alaska Hunting," *New York Times*, 24 Dec. 1972. On the MMPA's division of authority between NMFS and FWS, see Michael Bean and Melanie J. Rowland, *The Evolution of National Wildlife Law*, 3rd ed. (Westport, CT: Praeger, 1997), 109–147. On Hubbs's concerns about regulation, see Carl L. Hubbs to Scott McVay, 2 Feb. 1973, Box 54, Hubbs Papers.

17. Robert M. White, "Kenneth S. Norris, et al.," *Federal Register* 38 (24 Jan. 1973): 2340; Kenneth S. Norris to Hubbs, 7 June 1972, Norris to Hubbs, 20 Mar. 1973, Box 56, Hubbs Papers; *Marine Mammal Protection Oversight, Hearings before the Subcommittee on Fisheries and Wildlife Conservation and the Environment of the Committee on Merchant Marine and Fisheries, House of Representatives, on Oversight of the Marine Mammal Protection Act of 1972, La Jolla, Calif., August 21, 1973, Anchorage Alaska, August 31, 1973, Washington, D.C., January 16, 17, 1974*, 93rd Cong., 1st sess. (Washington, DC: U.S. Government Printing Office, 1974), on 154, 208, 442.

18. On Scheffer's background, see Victor B. Scheffer, "Exploring with Olaus Murie: The Aleutian Expedition," *Living Wilderness* 44 (June 1980): 11–17, on 12. The "naive idealist" quote is from J. L. McHugh to Carl L. Hubbs, 8 Dec. 1972, Box 16, Hubbs Papers.

19. G. Carleton Ray to Victor B. Scheffer, 1 Nov. 1973, Box 2, Entry 36, Office of Naval Research Records, Record Group 298, National Archives and Records Administration, College Park, MD; *Marine Mammal Protection Oversight Hearings, 1973–1974*, on 459. In 1975, HSUS awarded Scheffer its Joseph Wood Krutch Medal, named for the American nature writer and opponent of hunting; "Victor B. Scheffer Is the 1975 Recipient of the Joseph Wood Krutch Medal," *Marine Mammal News* 1 (Oct. 1975): 8.

20. Victor B. Scheffer to Gordon Gunter, 21 Sept. 1973, Box 17, Hubbs Papers.

21. John D. Dingell to Victor B. Scheffer, 9 Oct. 1973, in *Marine Mammal Protection Oversight Hearings, 1973–1974*, 471–472, on 472; Victor B. Scheffer to A. Starker Leopold, John J. Ryther, Lee M. Talbot, and Douglas G. Chapman and Dir., NMFS, and

Dir., BSFW, 2 Nov. 1973, in *Marine Mammal Protection Oversight Hearings, 1973–1974,* 469; Dingell to William E. Schevill, 7 Jan. 1974, in *Marine Mammal Protection Oversight Hearings, 1973–1974,* 468–469, on 469.

22. Frank M. Potter Jr. to Victor B. Scheffer, 2 Nov. 1973, in *Marine Mammal Protection Oversight Hearings, 1973–1974,* 467–468, on 468; *Marine Mammal Protection Oversight Hearings, 1973–1974,* 462, 463.

23. Christine Stevens to Robert Schoning, 11 Jan. 1974, in *Marine Mammal Protection Oversight Hearings, 1973–1974,* on 494; *Marine Mammal Protection Oversight Hearings, 1973–1974,* Scheffer's question about tagging on 439, Stevens's comments on marking on 408, exchange between Stevens and Potter on 416–417.

24. Jack W. Gehringer, "Sea World, Inc.: Notice of Application for Public Display Permit," *Federal Register* 24 (4 Feb. 1974): 4496; William E. Evans to Frank Powell, 21 May 1973, Box 34, Hubbs Papers.

25. Albert W. Erickson to Frank Powell, 30 Apr. 1974, Box 34, Hubbs Papers. The "madman" quote is from Siniff Interview. Erickson was one of the outside consultants, along with Frederick Dean and the Craigheads, who had been invited to the key September 1969 Natural Science Advisory Committee meeting on grizzly management in Yellowstone; Craighead, *Track of the Grizzly,* 203. On Erickson's early work with black bears, see A. W. Erickson, "The Age of Self-Sufficiency in the Black Bear," *Journal of Wildlife Management* 23 (1959): 401–405.

26. Robert Schoning, "Sea World, Inc.: Issuance of Permit for Marine Mammals," *Federal Register* 39 (13 May 1974): 17114; Albert W. Erickson, *Population Studies of Killer Whales (Orcinus orca) in the Pacific Northwest: A Radio-Marking and Tracking Study of Killer Whales* (Washington, DC: MMC, 1978), on 1; Robert F. Hutton, "Albert W. Erickson: Notice of Receipt of Application for a Scientific Research Permit," *Federal Register* 40 (8 Apr. 1975): 15925; Jack W. Gehringer, "Albert W. Erickson: Issuance of Permit for Marine Mammals," *Federal Register* 40 (18 June 1975): 25701.

27. *Marine Mammal Protection Oversight, Hearings before the Subcommittee on Fisheries and Wildlife Conservation and the Environment of the Committee on Merchant Marine and Fisheries, House of Representatives, on Oversight of the Marine Mammal Protection Act of 1972; To Review the Implementation, Administration, and Enforcement of the Act, October 21, 29, 30, December 9, 1975,* 94th Cong., 1st sess. (Washington, DC: U.S. Government Printing Office, 1976), on 6, 30, 153.

28. *Marine Mammal Protection Oversight Hearings, 1975,* 150.

29. *Marine Mammal Protection Oversight Hearings, 1975,* 205–207.

30. "Three-Way Confrontations among Animals, Their Protectors and Their 'Exploiters,'" *Marine Mammal News* 2 (Mar. 1976): 2; *Marine Mammal Amendments, Hearings before the Subcommittee on Fisheries and Wildlife Conservation and the Environment of the Committee on Merchant Marine and Fisheries, House of Representatives, On Save the Great Whales—H.J. Res. 923, April 30, 1976, Killer Whales—H.R. 12460, H.R. 12646, S. 3130, May 4, 1976, Tuna-Porpoise Amendments—H.R. 13865, H.R. 13883, May 20, 21, 24, 1976,* 94th Cong., 2nd sess. (Washington, DC: U.S. Government Printing Office, 1976), on 58.

31. *Congressional Record,* 11 Mar. 1976, S 6248–6249; *Congressional Record,* 29 Mar. 1976, S 8387–8395, on 8387, 8388, 8392.

32. Rex Weyler, *Greenpeace: How a Group of Journalists, Ecologists, and Visionaries Changed the World* (Vancouver, BC: Raincoast Books, 2004), 344–345; Alexandra Morton,

Listening to Whales: What the Orcas Have Taught Us (New York: Ballantine, 2002), 236; *Marine Mammal Amendments Hearings, 1976,* on 58–59, 79.

33. Carl Hubbs to Senators Cranston and Tunney and Representatives Hinshaw, Wilson, van Deerlin, and Burgener, telegram, 21 Mar. 1976, Box 34, Hubbs Papers; *Marine Mammal Amendments Hearings, 1976,* on 90–91, 107; "The Magnuson Bill to Ban All Taking of Killer Whales . . . ," *Marine Mammal News* 2 (May 1976): 4.

34. Erickson, *Population Studies,* iv, 1, 5, quote on 30; "Having Had 'Fair Success' with Radio-Tracking of Killer Whales . . . ," *Marine Mammal News* 2 (July 1976): 3–4. In August 1974, the NMFS also granted a permit to a team of NMFS marine mammalogists in Seattle to mark or radio-tag up to twenty-five killer whales, but the permit seems never to have been exercised; Jack W. Gehringer, "Northwest Fisheries Center: Issuance of Permit for Marine Mammals," *Federal Register* 39 (23 Aug. 1974): 30532.

35. Evans, *Fifty Years,* 91; Richard B. Lippin to Frank Powell Jr., 28 Nov. 1975, Box 22, Hubbs Papers.

36. "Biographical Sketch of William Eugene Evans," in *Hearing on Nomination of Dr. William Evans,* 13–14; "H-SWRI Scientific Advisory Council Meeting," 23 May 1977, Richard B. Lippin to Frank Powell Jr., 28 Nov. 1975, Box 22, Hubbs Papers; *Three Worlds of Sea World: Entertainment, Education, Research* (San Diego, CA: Sea World, 1977), Box 34, Hubbs Papers; Susan G. Davis, *Spectacular Nature: Corporate Culture and the Sea World Experience* (Berkeley: University of California Press, 1997), 69, 241–243.

37. Douglas G. Chapman, "Position Paper on Killer Whales in Puget Sound," unpublished manuscript, n.d., [1977], 13 pp., on 1, 16, Donald B. Siniff to Douglas Chapman, 12 May 1977, Box 11, Chapman Papers; Siniff to Chapman, 9 Sept. 1977, Robert Eisenbud to Chapman, 14 Aug. 1978, Box 9, Chapman Papers. On other marine mammalogists' skepticism toward photo-identification at this time, see Morton, *Listening to Whales,* 69.

38. On later identification of the two orcas tagged by Erickson, see Morton, *Listening to Whales,* 188. On the photo-identification technique, see Kenneth C. Balcomb III and Michael A. Bigg, "Population Biology of the Three Resident Killer Whale Pods in Puget Sound and Off Southern Vancouver Island," in Barbara C. Kirkevold and Joan S. Lockard, eds., *Behavioral Biology of Killer Whales* (New York: Alan R. Liss, 1986): 85–95; Morton, *Listening to Whales,* 142–143. On Bigg's abortive experiment with radio tagging, see Robin W. Baird, *Killer Whales of the World: Natural History and Conservation* (Stillwater, MN: Voyageur, 2002), on 86.

39. In 1978, Cornell, Sea World's chief veterinarian, submitted an application to NMFS to capture and import four killer whales from Icelandic or Scandinavian waters; "Sea World, Inc.: Receipt of Application for Permit," *Federal Register* 43 (2 May 1978): 18735. On the shift to Icelandic killer whales, see Sarah Lazarus, *Troubled Waters: The Changing Fortunes of Whales and Dolphins* (London: Collingwood, 2006), 159–160. On the Sealand capture, see "Orcas Face Capture off Canada," *Greenpeace Examiner* 7 (Summer 1982): 4–5. Morton, *Listening to Whales,* 148–149, 152–153; "Another Captive Orca Dies," *Greenpeace Examiner* 7 (Fall 1982): 18; Alan Reichman, "Sealand Death Toll Rises, Critics Demand End," *Greenpeace Examiner* 8 (Summer 1983): 18. For the masthead logo see *Greenpeace Chronicles* 1 (Autumn 1975): 1. On the design of the logo, see Weyler, *Greenpeace,* 376.

40. Roger S. Payne and Scott McVay, "Songs of Humpback Whales," *Science* 173

(1971): 585–597; Roger S. Payne, introduction to Payne, ed., *Communication and Behavior of Whales* (Boulder, CO: Westview, 1983): 1–6, on 1; Roger Payne et al., "External Features in Southern Right Whales (*Eubalaena australis*) and Their Use in Identifying Individuals," in Payne, *Communication*, 371–445, on 427, 429.

41. "What Do You Do with a Saved Whale?" *Greenpeace Examiner* 8 (Fall 1983): 4–5; Roger Payne, "Recently Developed Benign Research Techniques for Assessing and Modeling Whale Populations," paper prepared for the Global Conference on the Non-Consumptive Utilization of Cetacean Resources, Boston, MA, June 1983, 15 pp., on p. 1, Box 9, Chapman Papers; Robbins Barstow, "Non-Consumptive Utilization of Whales," *Ambio* 15 (1986): 155–163, on 157–158.

42. Dale Jamieson and Tom Regan, "Whales Are Not Cetacean Resources: An Animal Rights View," paper prepared for the Global Conference on Non-Consumptive Utilization of Cetacean Resources, Boston, MA, June 1983, 15 pp., on p. 14, in International Whaling Commission Archives, National Marine Mammal Laboratory, National Marine Fisheries Service, Seattle, WA. Regan's *The Case for Animal Rights*, published the year of the Whales Alive conference, provided a rights-based or deontological alternative to Peter Singer's utilitarian approach to animal ethics; Tom Regan, *The Case for Animal Rights* (Berkeley: University of California Press, 1983): Peter Singer, *Animal Liberation: A New Ethics for Our Treatment of Animals* (New York: Avon Books, 1977).

43. R. B. Brumsted, "National Marine Fisheries Service; Receipt of Application for Permit," 10 Mar. 1983, *Federal Register* 48 (17 Mar. 1983): 11310. The comparison to Sealand is in "Sea World Applies to Capture 100 'Specimens,'" *Greenpeace Examiner* 8 (Summer 1983): 19. Scheffer is quoted in "Sea World Orca Capture," *Greenpeace Examiner* 8 (Summer 1983): 11.

44. Richard B. Roe, "Notice of Hearing on Permit Application," 12 July 1983, *Federal Register* 48 (15 July 1983): 32376. Greenpeace's report on the hearing is in "Orcas' Fate on Hold," *Greenpeace Examiner* 8 (Fall 1983): 4. David Freed, "Plan to Capture Killer Whales Nets Anger," *Los Angeles Times*, 7 Aug. 1983; "For Knowledge and Pleasure," *Los Angeles Times*, 19 Sept. 1983.

45. *Hearing before the Subcommittee on Fisheries and Wildlife Conservation and the Environment of the Committee on Merchant Marine and Fisheries, House of Representatives, on Marine Mammal Protection Act Reauthorization, March 15, 1984*, 98th Cong., 2nd sess. (Washington, DC: U.S. Government Printing Office, 1984), 67–68; Carmen J. Blondin, "Marine Mammal Permits; Sea World, Inc.; Issuance," *Federal Register* 48 (4 Nov. 1983): 50915–50916; Mark Foster, "Sea World Wins OK to Capture Killer Whales," *Los Angeles Times*, 2 Nov. 1983.

46. "Protection of the Killer 'Orca' Whale," *Congressional Record*, 17 Nov. 1983, H 33284; *Hearing on MMPA Reauthorization, 1984*, on 59–60.

47. Norris's speech is quoted in "Looking Back at Ten Years of the Marine Mammal Protection Act . . . ," *Marine Mammal News* 9 (Dec. 1983): 1–2; see also Laura Tangley, "Beyond Bean Counting . . . : Marine Mammal Research Moves Ahead," *BioScience* 34 (Feb. 1984): 71–74. See also Evans, *Fifty Years*, 99–101.

48. "Kenneth Norris Made Headlines in New Zealand . . . ," *Marine Mammal News* 11 (Jan. 1985): 4–5; "Kenneth Norris Continues to Speak Out on Efforts by Greenpeace," *Marine Mammal News* 11 (Apr. 1985): 4; Kenneth Norris, "The Dolphin Connection," unpublished manuscript, n.d., [1985] Box 7, Chapman Papers.

49. Sidney Holt to R. Warnecke, 26 Feb. 1985, S. [Sidney Holt] to Ken Norris, 11 Mar. 1985, Box 10, Chapman Papers.

50. *Hearing on MMPA Reauthorization, 1984*, on 61–62.

51. Richard F. Ferraro and Merrill P. Spencer to Rod Chandler, 13 Mar. 1984, in *Hearing on MMPA Reauthorization, 1984*, on 182–184; *Hearing on MMPA Reauthorization, 1984*, on 74, 79, 98; Jim Hazelwood, "Speaking to Masters of the Sea," *Science News* 95 (18 Jan. 1969): 74–75, 77, on 77; Weyler, *Greenpeace*, 387–388. The MMC would use McCloskey's argument about the effect of capture operations on research in 1985 when it urged NMFS to seek ways to prevent the collection of dolphins off of southwestern Florida for public display from interfering with ongoing studies; *Annual Report of the Marine Mammal Commission, Calendar Year 1986: A Report to Congress* (Washington, DC: MMC, 1987), 145–146.

52. *Hearing on MMPA Reauthorization, 1984*, on 66, 90; *Hearing on Nomination of Dr. William Evans*, on 2, 6. Evans was the first chairman to be confirmed by the Senate, a requirement imposed in the previous reauthorization of the MMPA; "William Evans, Executive Director of Hubbs–Sea World in San Diego, CA, Is President Reagan's Choice to Chair the Marine Mammal Commission . . . ," *Marine Mammal News* 9 (Nov. 1983): 1–2. On Evans's recusal, see *Hearing on Nomination of Dr. William Evans*, on 3; William E. Evans to John R. Twiss Jr., 19 Mar. 1984, in *Hearing on Nomination of Dr. William Evans*, on 4–5.

53. *Hearing on Nomination of Dr. William Evans*, 7, 12–13; see also Rita Calvano, "Marine Mammal Panel Chief Sees No Conflict," *San Diego Union-Tribune*, 8 May 1984.

54. "Alaskans Act for Orcas," *Greenpeace Examiner* 9 (Spring 1984): 4. Press coverage of opposition to the capture often noted that opponents included the indigenous peoples of southeastern Alaska. For example, Ed Gamble, the Tlingit mayor of the town of Angoon, was quoted in the *Los Angeles Times* as saying that his people "don't like to mess with nature; we feel we have a right to live free, and the animals have that same right"; David Freed, "Indians Oppose Sea World: Killer Whale Capture Plan Riles Up Alaska," *Los Angeles Times*, 13 May 1984. On Sheffield's opposition, see David Freed, "Sea World Studying Decision: Alaska Harpoons Plan to Capture Killer Whales," *Los Angeles Times*, 26 May 1984. The Sierra Club Legal Defense Fund had been established in 1971 as a quasi-independent organization to provide legal services to the Sierra Club and other environmental organizations; it later became the fully independent organization Earthjustice; J. Michael McCloskey, *In the Thick of It: My Life in the Sierra Club* (Washington, DC: Island, 2005), 153. On the court case, see "Debate Derails SeaWorld [*sic*] Plan," *Greenpeace Examiner* 9 (Oct.–Dec. 1984): 8; David Freed, "Sea World Postpones Hunt for Whales," *Los Angeles Times*, 10 July 1984; David Freed, "Federal Judge Voids Sea World's Permit to Capture Whales," *Los Angeles Times*, 22 Jan. 1985; Jim Schachter, "Both Sides Say Ruling on Whales Was Victory," *Los Angeles Times*, 21 June 1986.

55. Richard B. Roe, "Marine Mammals; Application for Permit; Mr. A. Rus Hoelzel (P377)," 26 Mar. 1986, *Federal Register* 51 (2 Apr. 1986): 11334; Richard B. Roe, "Marine Mammals; Issuance of Permit; A. Rus Hoelzel," 22 Aug. 1986, *Federal Register* 51 (2 Sept. 1986): 31162–31163.

56. "Whale Lovers Quarreling: Biologist, Greenpeace Disagree about Research," *Spokesman-Review/Spokane Chronicle*, 21 Sept. 1986.

57. "Darting of Whales Target of Lawsuit," *Spokesman-Review*, 22 Oct. 1986; "Judge Bars 'Biopsy Dart' Tests on Orcas," *Seattle Post-Intelligencer*, 6 June 1987; *Annual Report of*

the MMC, 1986, 147. Hoelzel's eventual report of his thesis research, published in 1991, noted that "it was not possible to obtain large sample sizes" and thanked killer whale researchers and a number of marine parks, including Sea World, the Miami Seaquarium, the Vancouver Public Aquarium, and the Windsor Safari Park for providing tissue samples; A. Rus Hoelzel and Gabriel A. Dover, "Genetic Differentiation between Sympatric Killer Whale Populations," *Heredity* 66 (1991): 191–195, on 191.

58. Andrew Davis, "Can We Save Marine Mammals? The Deadly Decline of the Marine Mammal Protection Act," *Greenpeace* 1 (Jan.–Feb. 1988): 11–15, on 11, 14; "Back by Popular Demand, the MMPA," *Greenpeace Quarterly* 13 (May–June 1988): 12.

59. The MMC's account of the 1988 reauthorization process is in *Annual Report of the Marine Mammal Commission, Calendar Year 1989: A Report to Congress* (Washington, DC: MMC, 1990), 201. On Evans's nomination as director of NMFS, see "The Marine Mammal Commission Is Temporarily without One Commissioner," *Marine Mammals News* 12 (Sept. 1986): 1–2. Evans's exchange with Anderson is in *Hearing before the Subcommittee on Fisheries and Wildlife Conservation and the Environment, Committee on Merchant Marine and Fisheries, House of Representatives, on Reauthorization of the Marine Mammal Protection Act of 1972 . . . 10 May 1988*, 100th Cong., 2nd sess. (Washington, DC: U.S. Government Printing Office, 1988), 213–233, on 229; *Hearing on Reauthorization of the MMPA, 1988*, 29.

60. *Hearing on Reauthorization of the MMPA, 1988*, 4–5, 128, 130, 285–286.

61. *Annual Report of the MMC, 1989*, 11; "MMPA Reauthorized, Thanks to You," *Greenpeace* 14 (Jan.–Feb. 1989): 18–19.

62. *Annual Report of the MMC, 1989*, 202; Katherine Ralls and Robert L. Brownell Jr., "Protected Species: Research Permits and the Value of Basic Research," *BioScience* 39 (June 1989): 394–396, on 394, 396. Both Ralls and Brownell had personal experience with attempts by "the public" to influence their research practices. Ralls, for instance, had clashed with Friends of the Sea Otter, which was concerned about research that she and Siniff had proposed involving the use of surgically implanted radio tags to study California's southern sea otters; Donald B. Siniff and Katherine Ralls, "Reproduction, Survival and Tag Loss in California Sea Otters," *Marine Mammal Science* 7 (1991): 211–229; Rachel T. Saunders to Friends of the Sea Otter Advisory Committee, 6 Oct. 1986, Carton 2, Margaret Wentworth Owings Papers, MSS 86/71c, Bancroft Library, University of California, Berkeley.

63. The Friday Harbor workshop and report are described in "The Need to Use Telemetry and Biopsy Sampling to Learn More about Killer Whales . . . ," *Marine Mammal News* 16 (July 1990): 3. Robin William Baird, "Foraging Behaviour and Ecology of *Transient* Killer Whales (*Orcinus Orca*)," Ph.D. thesis, Simon Fraser University, 1994, 157 pp., on 49–50, 55, 64; "Jeff Goodyear, Cetacean-Rescue/Research, Inc., Trinity T.B., Newfoundland, Canada," *Marine Mammal Information*, Oregon State University, Sea Grant College Program, July 1982, 19. After his initial success in attaching time-depth recorders to killer whales using Goodyear's suction-cup design, Baird expanded his use of the technique to other killer whales in the Pacific Northwest as well as other regions and other species of small cetaceans, such as Dall's porpoises; Herbert W. Kaufman, "Marine Mammals," I.D. 040894D, *Federal Register* 59 (19 Apr. 1994): 18522; Herbert W. Kaufman, "Marine Mammals," I.D. 060694D, *Federal Register* 59 (17 June 1994): 31217; "Marine Mammals," I.D. 052196C, *Federal Register* 61 (31 May 1996): 27337–27338; "Marine Mammals; Permit No. 926 (P562)," I.D. 01239BA, *Federal Register* 63 (30 Jan. 1998): 4625–4626.

64. *Marine Mammal Commission Annual Report to Congress, 1992* (Washington, DC: MMC, 1993), 186–187.

65. *Hearing before the Committee on Commerce, Science, and Transportation on the Reauthorization of the Marine Mammal Protection Act, United States Senate, July 14, 1993,* 103rd Cong., 1st sess. (Washington, DC: U.S. Government Printing Office, 1993), on 1. As anthropologist Anne Brydon argues, *Free Willy* and its sequels presented redemptive stories of homecoming in which children learn "to trust . . . in the bonds of the modern nuclear family's love and strength," but they also reflected legal, institutional, and cultural changes that had taken place since the first killer whales were brought into captivity in the 1960s; Anne Brydon, "The Predicament of Nature: Keiko the Whale and the Cultural Politics of Whaling in Iceland," *Anthropological Quarterly* 79 (2006): 225–260, on 240.

66. *Hearing on the Reauthorization of the MMPA, 1993,* on 55.

67. *Hearing on the Reauthorization of the MMPA, 1993,* on 53, 57.

68. *Hearing on the Reauthorization of the MMPA, 1993,* on 18, 23, 24.

69. *Marine Mammal Commission Annual Report to Congress, 1995* (Washington, DC: MMC, 1996), 189–193; NMFS, *Marine Mammal Protection Act of 1972 Annual Report, January 1, 1994 to December 31, 1994,* on 6; Gary Brone et al., "Permit Programs," in NMFS, *Marine Mammal Protection Act of 1972 Annual Report, January 1, 1995 to December 31, 1995,* 82.

70. Kenneth Brower, *Freeing Keiko: The Journey of a Killer Whale from Free Willy to the Wild* (New York: Gotham Books, 2005), 33, 35.

71. Morton, *Listening to Whales,* 236; Brower, *Freeing Keiko,* 37; M. Simon et al., "From Captivity to the Wild and Back: An Attempt to Release Keiko the Whale," *Marine Mammal Science* 25 (2009): 693–705, on 694; Ann D. Terbush, "Marine Mammals," I.D. 080295C, 3 Aug. 1995, *Federal Register* 60 (14 Aug. 1995), 41881; Ann D. Terbush, "Marine Mammals," I.D. 091495B, 22 Sept. 1995, *Federal Register* 60 (29 Sept. 1995), on 50555.

72. For various NMFS and MMC statements of the need for a release permit, see Brone et al., "Permit Programs," 83; Ted Bardacke, "Willy or Won't He? In the Movie, the Whale Swims Free. For the Ailing Keiko, It's Not So Simple," *Washington Post,* 2 Oct. 1994; *Marine Mammal Commission Annual Report to Congress, 1998* (Washington, DC: MMC, 1999), 201–206; NMFS, *Marine Mammal Protection Act of 1972 Annual Report, January 1, 1997 to December 31, 1997,* 121.

73. NMFS, *Marine Mammal Protection Act of 1972 Annual Report, January 1, 1996 to December 31, 1996,* on 82–83; NMFS, *Administration of the Marine Mammal Protection Act of 1972, Annual Report, 1999–2000,* 48; NMFS, "What Should We Know before We Free Willy?" PBS Online, May–June 1996, www.pbs.org/wgbh/pages/frontline.

74. "Marine Mammals in Captivity," in *MMC Annual Report, 1998,* 201–206, on 204; *Free Willy 3: The Rescue* (1997), directed by Sam Pillsbury.

75. NMFS, *Marine Mammal Protection Act of 1972 Annual Report, January 1, 1998 to December 31, 1998,* 71, 82; "Permit Programs," in NMFS, *MMPA Annual Report, 1997,* 115–122, on 121; Katy Muldoon, "Lack of Data, Questions of Ethics Muddy Waters on Keiko's Future," *Oregonian,* 2 Nov. 1997; *MMC Annual Report, 1998,* 206. The export of Keiko did not require a permit under the MMPA, but it did require U.S. Department of Agriculture and NMFS approval. NMFS had asserted, with support from the MMC, its right to compel on-site inspections of foreign facilities to ensure that they met stan-

dards, but USDA had argued that written descriptions were sufficient. In August 1997, NMFS told the MMC that it believed it had the statutory authority to demand "letters of comity"—agreements to enforce NMFS's decisions—from foreign governments as a condition of receiving marine mammals. In theory, such a letter would give NMFS authority over the foreign facility for as long as it held the exported animal, but whether such letters had any authority was a subject of much debate; *MMC Annual Report, 1995*, on 207; *MMC Annual Report, 1998*, 201–204.

76. Katy Muldoon, "Thanks to Tracking Device, Keiko Would Never Be Alone in the Wild," *Oregonian*, 3 Aug. 1999.

77. Morris Bradley Hanson, "An Evaluation of the Relationship between Small Cetacean Tag Design and Attachment Durations: A Bioengineering Approach," Ph.D. thesis, University of Washington, 2001, 209 pp., on vi; Brad Hanson, "Harbor Porpoise Tagging Research in the San Juan Islands," NOAA-NMFS NMML, Quarterly Report for July-Aug.-Sept. 1998, www.afsc.noaa.gov/Quarterly; Rich Ferrero, "Tagging of Beluga Whale in Cook Inlet, Alaska," NOAA-NMFS NMML, Quarterly Report for July-Aug.-Sept. 1999, www.afsc.noaa.gov/Quarterly; Jeff Cooke, "Dall's Porpoise Tagging," NOAA-NMFS NMML, Quarterly Report for July-Aug.-Sept. 1999, www.afsc.noaa.gov/Quarterly; NMFS, *Marine Mammal Protection Act Annual Report, 1999–2000*, 47; Brad Hanson, "Cetacean Tagging," NOAA-NMFS NMML, Quarterly Report for Apr.-May-June 2000, www.afsc.noaa.gov/Quarterly.

78. Brower, *Freeing Keiko*, 217–218; "One Giant Step for Keiko," *Washington Post*, 26 May 2000; *Marine Mammal Commission Annual Report to Congress, 2000* (Washington, DC: MMC, 2001), 217; Katy Muldoon, "Powerful Icelandic Earthquakes Fail to Shake Unflappable Keiko," *Oregonian*, 22 June 2000.

79. *MMC Annual Report, 2000*, 218; Brower, *Freeing Keiko*, on 211; Donald G. McNeil Jr., "Keiko Makes It Clear: His 'Free Willy' Was Just a Role," *New York Times*, 6 Nov. 2001; *Marine Mammal Commission Annual Report to Congress, 2001* (Washington, DC: MMC, 2002), 212.

80. Katy Muldoon, "Keiko Lands New Deep-Pocket Friend," *Oregonian*, 3 Aug. 2002; HSUS, "Keiko Spends Unprecedented 41 Days in Wild; HSUS Says Quest for Freedom May Be Fulfilled," 16 Aug. 2002, http://www.hsus.org/press_and_publications/press_releases/keiko_spends_unprecedented_41_days_in_wild_hsus_says_quest_for_freedom_may_be_fulfilled.html; Simon et al., "From Captivity to the Wild and Back," 2; Brower, *Freeing Keiko*, on 250; HSUS, "Keiko Takes an Unprecedented Walk in the North Atlantic," 2 Aug. 2002, http://www.hsus.org/marine_mammals/marine_mammals_news/keiko_watch_keiko_takes_an_unprecedented_walk_in_the_north_atlantic_822002.html; HSUS, "Data Indicate Keiko May Be Eating on His Own," 16 Aug. 2002, http://www.hsus.org/marine_mammals/marine_mammals_news/keiko_watch_data_indicate_keiko_may_be_eating_on_his_own_8162002.html.

81. HSUS, "Keiko Takes an Unprecedented Walk"; HSUS, "Keiko Spends Unprecedented 41 Days in Wild"; Brower, *Freeing Keiko*, 260.

82. Katy Muldoon, "Ex-Caregivers Challenge Report on Keiko's Health," *Oregonian*, 23 Aug. 2002; *Marine Mammal Commission Annual Report to Congress, 2002* (Washington, DC: MMC, 2003), 219–220.

83. Brower, *Freeing Keiko*, 262; Katy Muldoon, "Seaquarium Makes a Play to Own Keiko," *Oregonian*, 21 Sept. 2002; Katy Muldoon, "Seaquarium's Bid for Keiko Denied," *Oregonian*, 4 Oct. 2002; HSUS, "Keiko Watch: Authorities Rebuff Miami Seaquarium's

Request," 11 Oct. 2002, www.hsus.org/marine_mammals/marine_mammals_news; HSUS, "Keiko, the Most Famous Whale in the World, Dies in Norway," 12 Dec. 2003, www.hsus.org/press_and_publications/press_releases; Vinick quoted in Sandi Doughton, "Keiko Won Hearts on Screen, in Real Life," *Seattle Times*, 13 Dec. 2003.

84. An influential recent formulation of this vision is William Cronon, "The Trouble with Wilderness; or, Getting Back to the Wrong Nature," *Environmental History* 1 (1996): 7–28. On the listing of the southern resident killer whale population under the ESA, see NMFS, *Recovery Plan for the Southern Resident Killer Whales* (Orcinus orca) (Seattle: NMFS, Northwest Region, 2008), V-39.

CONCLUSION: New Connections

1. The first successful satellite telemetry study of birds was conducted on wandering albatrosses in the southern Indian ocean in 1989; Pierre Jouventin and Henri Weimerskirch, "Satellite Tracking of Wandering Albatrosses," *Nature* 343 (1990): 746–748. Anderson's NSF award is briefly described in "CAREER: Evolutionary Ecology of Avian Reproduction," Award Abstract 9629539, in NSF's Award Search database, www.nsf.gov/awardsearch. For the invitation to the public to participate in the Albatross Project, see Scott Lafee, "Public Has Opportunity to Help Science Take Flight," *San Diego Union-Tribune*, 18 Feb. 1998.

2. Margaret Williams et al., "Satellite Tracking of Threatened Species," *Argos Newsletter* 53 (Aug. 1998), www.cls.fr/html/argos/documents; Wendee Holtcamp, "Tracking by Internet," *Audubon* 100 (1 Sept. 1998): 57; Lafee, "Public Has Opportunity to Help Science."

3. Carl Safina, *Eye of the Albatross: Visions of Hope and Survival* (New York: Holt, 2002), on xiii, 97–98.

4. J. P. Croxall, "Research and Conservation: A Future for Albatrosses?" in Graham Robertson and Rosemary Gales, eds., *Albatross Biology and Conservation* (Chipping Norton, Australia: Beaty, 1997): 276–286, on 286; P. A. Prince, A. G. Wood, T. Barton, and J. P. Croxall, "Satellite Tracking of Wandering Albatrosses (Diomedea Exulans) in the South Atlantic," *Antarctic Science* 4 (1992): 31–36; Ladbrokes, "Big Bird Race to Highlight Plight of the Albatross," press release, 22 Jan. 2004, www.ladbrokes.com; see also http://betting.ladbrokes.com/en/big-bird-race.

5. On Crittercam, see www.nationalgeographic.com/crittercam. On the animal-borne imaging symposium, see www.nationalgeographic.com/abis. On the history of wildlife photography and film in the United States and the United Kingdom, see Gregg Mitman, *Reel Nature: America's Romance with Wildlife on Film* (Cambridge, MA: Harvard University Press, 1999); Derek Bousé, *Wildlife Films* (Philadelphia: University of Pennsylvania Press, 2000); Cynthia Chris, *Watching Wildlife* (Minneapolis: University of Minnesota Press, 2006); Donna Haraway, "Teddy Bear Patriarchy: Taxidermy in the Garden of Eden, New York City, 1908–1936," *Social Text* 11 (1984): 20–64. On the realistic wild animal story, see Ralph H. Lutts, *Nature Fakers: Wildlife, Science and Sentiment* (Golden, CO: Fulcrum, 1990); Ralph H. Lutts, ed., *The Wild Animal Story* (Philadelphia: Temple University Press, 1998). Donna Haraway argues that the *Crittercam* series' frank depiction of "the material-semiotic requirements of getting on together in specific lifeworlds" for scientists and the animals they tag makes it clear "that situated human beings have

epistemological-ethical obligations to the animals"; Haraway, "Crittercam," in *When Species Meet* (Minneapolis: University of Minnesota Press, 2008), 249–263, on 262–263.

6. Dwain W. Warner, "Preliminary Request for Research Grant and for Instrument Space on a Satellite in Polar Orbit," Grant Proposal to NASA, 16 April 1962, Warner Papers; Dwain W. Warner, "Space Tracks: Bioelectronics Extends Its Frontiers," *Natural History* 62 (1963): 8–15; Warner Interview; Cochran Interview; Cochran and Swenson Interview; Martin Wikelski et al., "Going Wild: What a Global Small-Animal Tracking System Could Do for Experimental Biologists," *Journal of Experimental Biology* 210 (2007): 181–186, on 181, 183, 185.

Essay on Sources

The technical development of wildlife radiotelemetry in the 1960s can be tracked through two periodicals: the *Journal of Wildlife Management*, which is widely available, and the Wildlife Society's *Wildlife Telemetry Newsletter* (1961–1967), which can be found in the library of the University of Minnesota and a few other locations. The proceedings of the 1962 biotelemetry conference at the American Museum of Natural History in New York is also useful, though its inclusion of both wildlife biology and laboratory-based physiological research gives an exaggerated sense of the connections between the two (Lloyd E. Slater, ed., *Bio-Telemetry: The Use of Telemetry in Animal Behavior and Physiology in Relation to Ecological Problems*, New York: Pergamon, 1963). Many of the participants in that conference also participated in a follow-up conference sponsored by the American Institute of Biological Sciences, the proceedings of which were published in a special issue of *BioScience* in 1965.

For details on the politics and practice of fieldwork, archival materials are essential. The William H. Marshall Papers at the University of Minnesota Archives in St. Paul, which were unprocessed when I consulted them in 2004, contain rich details on the Grousar project and, more generally, on the trials and tribulations of academic wildlife research from the 1940s to the 1970s as seen through Marshall's eyes. Records of the Cedar Creek project are more scattered, though a small collection of relevant records is available at the University of Minnesota Archives. A. C. Hodson's *History of the Cedar Creek Natural History Area* (St. Paul: University of Minnesota Field Biology Program, 1985) is more a compilation of primary source materials than it is a synthetic history; as such it provided some invaluable source material. Clarence L. Lehman allowed me unfettered access to old filing cabinets at the Cedar Creek laboratory building. Donald B. Siniff generously allowed me access to his personal papers and provided otherwise unavailable information about Cedar Creek in an interview.

Much of the history of the Cedar Creek automatic radio-tracking system and other aspects of the early development of radio tracking only became clear through the course of interviews with Siniff, John Tester, Larry Kuechle, William Cochran, George Swenson, David Mech, and Dwain Warner. Like Siniff, Warner and Cochran also shared documentary records from their personal collections with me. A visit to the library and photograph archives of the Illinois Natural History Survey provided useful information about Cochran and Lord's "radio rabbit" and duck-tracking work, as did an e-mail exchange with Lord, who generously sent me an unpublished manuscript describing the work. The records of the Office of Naval Research and the U.S. Fish and Wildlife Service at the National Archives in College Park, Maryland, contain a small amount of scattered but useful information about early wildlife radio-tracking efforts, and the records of the Minnesota Department of Conservation and the Minneapolis-Honeywell Regulator Company at the Minnesota Historical Society in St. Paul similarly shed some small light on the situation in Minnesota.

The Craigheads' own writings in *National Geographic* magazine and in scientific journals provide the best starting point for understanding their work on the grizzly bears of Yellowstone, along with Frank Craighead's 1979 account, *Track of the Grizzly* (San Francisco: Sierra Club Books, 1979), which is especially valuable for its one-sided but informative final chapter, titled "Bureaucracy and Bear." Because of its importance in the history of wildlife management and scientific research in the American national park system, the Craigheads' work in Yellowstone is unusually well described in the secondary literature and is documented in great detail in the archival records of the National Park Service at the National Archives in College Park and at the Yellowstone Heritage and Research Center in Gardiner, Montana, located just outside the park's northern entrance. Three books provided guidance in navigating these records: Richard Sellars's *Preserving Nature in the National Parks: A History* (New Haven, CT: Yale University Press, 1997), James Pritchard's *Preserving Yellowstone's Natural Conditions: Science and the Perception of Nature* (Lincoln: University of Nebraska Press, 1999), and Alice Wondrak Biel's *Do (Not) Feed the Bears: The Fitful History of Wildlife and Tourists in Yellowstone* (Lawrence: University of Kansas Press, 2006), though only the latter takes criticism of the Craigheads' research methods seriously. Further details of the Craigheads' work can be found in the quarterly and annual reports of the Montana Cooperative Wildlife Research Unit, available at the library of Montana State University and in the archival records of the Department of Zoology at the University of Montana.

Additional material on the grizzly controversy and the broader question of independent research in the national parks can be found in the papers of A. Starker Leopold at the Bancroft Library of the University of California, Berkeley, and in the papers of Carl L. Hubbs at the Scripps Institution of Oceanography in San Diego, California, who happened to be a member of the National Academy of Science committee on science in the national parks in 1963. Records on proposed grizzly studies in Mount McKinley National Park (later expanded and renamed Denali National Park and Preserve) were more difficult to find, in part because it is a story of a study that never took place. The Murie Family Papers at the American Heritage Center at the University of Wyoming in Laramie and the Olaus Murie Papers at the Denver Public Library helped me reconstruct the perspective of the Murie brothers and provided some insight into the positions of Park Service leadership. The records of the Wilderness Society in the Denver Public Library's conservation collection contained a smattering of relevant correspondence, as did the David Brower and Sierra Club records at the Bancroft Library. The "Monique" incident is documented extensively in the papers of Helmut Buechner and other collections at the Smithsonian Institution Archives in Washington, DC; Charles Cote generously helped contextualize her death and NASA's response to it in an interview in his office at NASA Goddard Space Flight Center.

Further information about research and management policies in Mount McKinley and about Will Troyer's caribou study, in particular, was found in the National Archives repository in Anchorage, Alaska, and in the libraries of the University of Alaska at Fairbanks and Anchorage. Visits to interpretive facilities at Yellowstone and, especially, Denali were essential for understanding the way the Park Service has presented this contested history to visitors. Interviews with John C. Seidensticker and Frederick C. Dean helped clarify many points.

The interview with Seidensticker was also crucial to understanding how the radio-tracking techniques that were developed for grizzlies in Yellowstone and in Seidensticker's subsequent work with mountain lions in Idaho were adapted for use on the tigers of Nepal's Royal Chitwan National Park. Hemanta Mishra was also generous with his time, providing not only an interview and a hot cup of tea at his home in northern Virginia but a draft of his then-unpublished book, *Bones of the Tiger: Protecting the Man-Eaters of Nepal* (Guilford, CT: Lyons, 2010). Mishra's *The Soul of a Rhino: A Nepali Adventure with Kings and Elephant Drivers, Billionaires, and Bureaucrats, Shamans and Scientists, and the Indian Rhinoceros* (Guilford, CT: Lyons, 2008), coauthored with Jim Ottaway Jr., provided information about Smithsonian's work in Nepal and insights into the complexities of cross-cultural collaborations, as did Eric Dinerstein's *Tigerland and Other Destinations* (Washington: Island, 2005). Siniff's personal papers provided additional insight into the conflict between the Smithsonian-Nepal Tiger Ecology Project and its antagonists in Chitwan from his perspective as an academic adviser to Melvin Sunquist and James L. David Smith. Fiona and Mel Sunquist's *Tiger Moon* (Chicago: University of Chicago Press, 1988), an account of their time in Chitwan, was also helpful.

The records of the World Wildlife Fund are not readily available, and its annual reports and other official publications provide only the slightest glimpse of the fascinating "diplomatic and political subtleties," as Sidney Galler described them, that accompany virtually all international wildlife conservation and research efforts. However, much of the history of WWF-US and its relations with WWF International and the International Union for the Conservation of Nature can be reconstructed through the papers of its board members, including those of Russell E. Train at the Library of Congress in Washington, DC, which were unprocessed when I consulted them but well organized by Train himself, and Harold J. Coolidge at the Harvard University Archives in Cambridge, Massachusetts.

By far the most important source for the history of the Smithsonian's involvement in Nepal and, more generally, for the history of American international field biology are the Smithsonian Archives, which are remarkably comprehensive and open. In navigating these archives I am indebted to Michael Lewis, whose *Inventing Global Ecology: Tracking the Biodiversity Ideal in India, 1945–1997* (Hyderabad, India: Orient Longman, 2003) used them to provide an insightful account of the Smithsonian's troubled history of collaborative field research in India, and to Pamela Henson, director of the Smithsonian's institutional history division. The historical blog posts that Christen Wemmer has interspersed with his fascinating photographs of northern California wildlife at Camera Trap Codger (cameratrapcodger.blogspot.com) clarified several key points.

No single archive provided the wealth of data for the history of killer whale radio tagging that the Smithsonian Archives provided for tigers, but the Hubbs Papers provided a broad view of the links between government, academia, the military, and the public display industry in marine mammal research through the 1970s, including much incidental material on the career of his student Kenneth Norris. The Hubbs papers include scattered but very illuminating material about Sea World, Marineland of the Pacific, and other marine parks, whose history remains obscure despite the important work of Susan Davis in her book *Spectacular Nature: Corporate Culture and the Sea World Experience*

(Berkeley: University of California Press, 1997) and of Gregg Mitman in his account of Marine Studios in *Reel Nature: America's Romance with Wildlife on Film* (Cambridge, MA: Harvard University Press, 1999). William Evans's memoir, *Fifty Years of Flukes and Flippers: A Little History and Personal Adventures with Dolphins, Whales and Sea Lions (1958–2007)* (Sofia: Pensoft, 2008), provided a broad-ranging view of marine mammal research from the 1950s to the 1990s, though one might wish that it had more information about the navy's secretive and understudied marine mammal program, about which Evans knew as much as anyone.

Detailed information about marine mammal research in progress can be found in *Marine Mammal Information*, a newsletter published by Bruce Mate at Oregon State University from 1978 to 1986, when its function was taken over by the publications of the Society for Marine Mammalogy. Another newsletter, the *Marine Mammal News*, provides strong coverage of policy developments in Washington. The library of the National Marine Fisheries Service's Southwest Fisheries Science Center, just up the road from Scripps, provided access to NMFS research progress reports and other hard-to-find literature, as did the Marine Mammal Laboratory at the Smithsonian's National Museum of Natural History in Washington, DC, the library of the Woods Hole Oceanographic Institution in Woods Hole, Massachusetts, and the library of NMFS's National Marine Mammal Laboratory in Seattle.

The latter's collection of reprints contained some hidden gems, and its archive of International Whaling Commission papers made that organization's involvement in whale tagging easier to track than it would have been otherwise. Because of Douglas Chapman's long-running involvement with the scientific committee of the IWC and the Committee of Scientific Advisors of the Marine Mammal Commission, his papers at the University of Washington Archives in Seattle are essential for anyone who wants to understand the role of science in these institutions or in modern whale conservation generally. The records of the National Science Foundation at the National Archives in College Park provided some useful information about that agency's funding for marine mammal research in the Antarctic, while the Office of Naval Research records there provided similar material on Arctic research.

The bible of American wildlife law, which includes a thorough discussion of the Marine Mammal Protection Act, is Michael J. Bean and Melanie J. Rowland, *The Evolution of National Wildlife Law* (Westport, CT: Praeger, 1997). While I did not conduct formal interviews with Jeff Foster, Brad Hanson, Robin Baird, or Naomi Rose, they generously answered my questions about radio tagging of killer whales, the campaign to free Keiko, and the complexities of the MMPA's permitting process through informal phone conservations and e-mail exchanges. I am also grateful to Greg Marshall for his help in untangling the history of the "remora" tag.

Beyond the specific sources listed above, two broad, partially overlapping categories of secondary literature provided the framework within which this study was conducted. The first regards the history of wildlife conservation and research. Arthur McEvoy's *The Fisherman's Problem: Ecology and Law in the California Fisheries, 1850–1980* (Cambridge: Cambridge University Press, 1986), Louis Warren's *The Hunter's Game: Poachers and Conservationists in Twentieth-Century America* (New Haven, CT: Yale University Press, 1997), and Karl Jacoby's *Crimes against Nature: Squatters, Poachers, Thieves, and the Hidden History of American Conservation* (Berkeley: University of California Press, 2001) show how Americans who hunted wild animals for a living became increasingly involved with state and national governments over the course of the twentieth century. Mark Barrow's *A Passion for Birds: American Ornithology after Audubon* is an illuminating case study of the tensions between scientists and amateurs and between research and conservation; his *Nature's Ghosts: Confronting Extinction from the Age of Jefferson to the Age of Ecology* (Chicago: University of Chicago Press, 2009) is essential reading for anyone interested in the history of extinction. Thomas Dunlap's *Saving America's Wildlife* (Princeton, NJ: Princeton University Press, 1988) provides a useful overview of changes in twentieth-century predator control policies; complementary accounts can be found in Christian Young's *In the Absence of Predators: Conservation and Controversy on the Kaibab Plateau* (Lincoln: University of Nebraska Press, 2002) and Timothy Rawson's *Changing Tracks: Predators and Politics in Mt. McKinley National Park* (Fairbanks: University of Alaska Press, 2001).

Writings on wilderness are too voluminous to even begin cataloging, but a good start is *The Great New Wilderness Debate*, edited by J. Baird Callicott and Michael P. Nelson (Athens: University of Georgia Press, 1998), which includes a reprint of William Cronon's essay "The Trouble with Wilderness" and many responses to it. Julianne Lutz Newton's *Aldo Leopold's Odyssey* (Washington, DC: Island / Shearwater Books, 2006), which builds on earlier biographies by Curt Meine and Susan Flader, provides the best extant account of the evolving thought of that founder of wildlife management. Gregg Mitman's *Reel Nature* provides an overview of the major shifts in American attitudes toward and motivations for wildlife conservation in the twentieth century; it can profitably be read alongside Cynthia Chris's *Watching Wildlife* (Minneapolis: University of Minnesota Press, 2006) and Derek Bousé's *Wildlife Films* (Philadelphia: University of Pennsylvania Press, 2000). Donna Haraway's *Primate Visions: Gender, Race, and Nature in the World of Modern Science* (New York: Routledge, 1989) is a provocative and insightful account of twentieth-century primatology, much of relevant to wildlife research of all kinds.

Although wildlife biology has sometimes been only tenuously connected to ecology, histories of ecology provide useful context, especially Donald Worster's *Nature's Economy: A History of Ecological Ideas* (New York: Cambridge University Press, 1994), whose broad temporal sweep and ambitious schematization remain unparalleled, even as it has been insightfully critiqued in works such as Peder Anker's *Imperial Ecology: Environmental Order in the British Empire, 1895–1945* (Cambridge, MA: Harvard University Press, 2001). Despite some problematic assumptions about disciplinary borders, Robert Kohler's *Landscapes and Labscapes: Exploring the Lab-Field Border in Biology* (Chicago: University of Chicago Press, 2002) raises important questions about the relationship between laboratory and field, and his *All Creatures: Naturalists, Collectors, and Biodiversity, 1850–1950* (New York: Princeton University Press, 2006) describes patterns of nature study established around the turn of the century that would continue to influence biologists in the late twentieth century. Stephen Bocking's *Ecologists and Environmental Politics: A History of Contemporary Ecology* (New Haven, CT: Yale University Press, 1997) reveals the essential role played by the Atomic Energy Commission in pushing ecosystem ecology to the forefront of the discipline after the Second World War.

Much remains to be done on the history of international wildlife conservation and research. John MacKenzie's *The Empire of Nature: Hunting, Conservation, and British Imperialism* (New York: Manchester University Press, 1988) is essential for understanding the colonial legacies that live on in postcolonial conservation, while Jane Carruthers's *The Kruger National Park: A Social and Political History* (Pietermaritzburg: University of Natal Press, 1995) documents the important South African case. Some of the most exciting recent work in this field, however, has been done not by historians but by anthropologists, geographers, and others working in political ecology, such as Nancy Peluso, Cori Hayden, Elizabeth Garland, Celia Lowe, Roderick Neumann, and Christine Walley.

The second category of secondary literature involves the relationship between science and its publics. Though they focus on Britain during the Victorian era, Harriet Ritvo's *The Animal Estate: The English and Other Creatures in the Victorian Age* (Cambridge, MA: Harvard University Press, 1987) and *The Platypus and the Mermaid, and Other Figments of the Classifying Imagination* (Cambridge, MA: Harvard University Press, 1997) present arguments about the vitality and diversity of vernacular understanding of animals and the natural world and the difficulty of neatly separating those understandings from "science" proper that have much broader relevance.

This book's guiding assumption that the quotidian instruments and practices of scientists can be the subjects and even the means of a rich political and cultural discourse is drawn from a number of works in science studies, notably Bruno Latour and Steve Woolgar's *Laboratory Life: The Construction of Scientific Facts* (Princeton, NJ: Princeton University Press, 1986), Latour's *The Pasteurization of France* (Cambridge, MA: Harvard University Press, 1988), Steven Shapin and Simon Schaffer's *Leviathan and the Air-Pump: Hobbes, Boyle, and the Experimental Life* (Princeton, NJ: Princeton University Press, 1985), Robert Kohler's *Lords of the Fly: Drosophila Genetics and the Experimental Life* (Chicago: University of Chicago Press, 1994), and Peter Galison's *Image and Logic: A Material Culture of Microphysics* (Chicago: University of Chicago Press, 1997).

This book also draws inspiration from the work of scholars who have sought to broaden the focus of science studies beyond the sites of research to include courtrooms, legislative hearings, popular media, and so forth, such as Chandra Mukerji's *A Fragile Power: Scientists and the State* (Princeton, NJ: Princeton University Press, 1989), Sheila Jasanoff's *The Fifth Branch: Science Advisers as Policymakers* (Cambridge, MA: Harvard University Press, 1990) and *Designs on Nature: Science and Democracy in Europe and the United States* (Princeton, NJ: Princeton University Press, 2005), Haraway's *Modest_Witness@Second_Millenium.FemaleMan©_Meets_OncoMouse™: Feminism and Technoscience* (New York: Routledge, 1997), and Stefan Helmreich's inspiring *Alien Ocean: An Anthropology of Marine Biology and the Limits of Life* (Berkeley: University of California Press, 2009).

Index